NAVAL MINE WARFARE MARKETS

Author: Anthony J Watts

January 1997

ISBN: 07106 1503 5

Sentinel House, 163 Brighton Road
Coulsdon, Surrey CR5 2NH, UK
Telephone: +44 (0)181 700 3700

In the USA and its dependencies:
Jane's Information Group Inc
1340 Braddock Place, Suite 300
Alexandria VA 22314-1651, United States of America
Telephone: +1 (703) 683 3700

A Jane's Special Report from Jane's Information Group

The US Avenger class mine countermeasures vessels (l to r) Gladiator, Warrior and Pioneer
(H M Steele)

CONTENTS

NAVAL MINE WARFARE MARKETS

FOREWORD

Development of *Naval Mine Warfare Markets* is the latest in our growing list of Jane's Special Reports which we are developing to add to the range of information services which we provide for the Defence and Aerospace communities.

The importance of mine countermeasures in the overall strategic considerations of navies can hardly be over–emphasised. The threat to major warships and their crews from the relatively cheap mine is very real and even relatively simple mines can, and do, still pose a significant danger to the technically sophisticated modern warship. To the layman it is of great interest that only relatively few navies have invested in and developed a fully operational and effective MCMV squadron. This Special Report provides a blue print to those initiating or developing their Mine Countermeasures capability – it offers to professionals involved, both in the Navies of the world and the Defence Industries who supply them, the independent and impartial thoughts and guidance that they expect from Jane's.

We have been fortunate to be able to persuade Tony Watts to prepare this report. He is a very experienced journalist and writer having worked for many years on naval matters, especially in the fields of underwater warfare and mine countermeasures. For the last seven years Tony has been the Editor of the respected Jane's yearbook *Jane's Underwater Warfare Systems*.

We trust that the reader will find this product of interest and value – naturally, all suggestions for change, additions or improvement will be most welcome as we plan to bring out further editions in future, assuming we get the expected encouragement from the market.

Jane's would also be pleased to offer consultancy services to support specific reader requirements.

K. E. Faulkner

K. E. Faulkner – Managing Editor, Naval and Maritime Group **20 December 1996**

A UK Sandown class minehunter *(H M Steele)*

EXECUTIVE SUMMARY

As the general world situation exhibits increasing instability, with the dominance of major power groups lessening, the need for adequate national defence forces to protect sovereign territory and interests will increase. In this context the mine countermeasures vessel is a vital element for any country seeking to deploy a balanced naval force in support of national interests.

The outcome is likely to be an increase in the number of navies seeking to develop MCM forces over the next 10 to 20 years as the mine threat becomes more powerful and insidious. As this report shows growth is most likely to occur in the Far East, with the Gulf and Western Asia following close behind.

Elsewhere in the world MCM forces are likely to remain more or less at their present strength. However, one region which may actually witness a decline in the strength of MCM forces is Europe. This will largely be accounted for by the scrapping by Russia of large numbers of obsolete mine warfare vessels. Former Warsaw Pact countries operating mine warfare vessels transferred by the former Soviet Union will also account for a reduction in MCM forces.

Of course any potential to modernise, expand or even develop a MCM force is dependent on funding. In the light of the Post Communist order in the West defence budgets have been severely pruned and military forces have been subject to critical re–appraisal as to their *modus operandi*. As a result some NATO countries are now actively considering reducing the strength of their MCM forces. The market in Europe is therefore somewhat limited at present, as this report shows. However, even if forces are reduced there will come a time around the turn of the century when some navies will have to give due consideration to replacing obsolete vessels if they are to maintain a credible MCM force.

The Far East, on the other hand, is a region which exhibits one of the fastest growing economies in the world. It is also a region which is showing increasing instability. The outcome is a desire on the part of a number of nations to either build up MCM capability, or modernise and expand an existing one. From a market viewpoint Australia is making a very strong endeavour to take on the mantle of main supplier to the region. However, to achieve this it will have to break the hold of European nations which have traditionally supplied into this market. South Korea with a burgeoning economy, expanding warship construction industry and well established civil electronics industry, which is now broadening its base to encompass military requirements, may well prove to be a major market force in the MCM field in about 10 to 15 years time.

South America and the Caribbean is a region in the grip of economic uncertainty and beset by budgetary problems. Funding new military projects, particularly high cost naval ones, is fraught with difficulties. Even agreeing funding for upgrading and modernisation is subject to constant delay and sometimes cancellation. The acquisition of second–hand hulls is also subject to intense negotiation with regard to the very limited funding available. With all their other problems, navies in the region do not, therefore, readily find themselves in a position to make adequate provision for MCM.

MCM will continue to lag behind the development of the mine. The gap is narrowing, however, and there are limits to the extent to which mines can be developed. The two main developments in the future will be the self–propelled mine and mines which can be laid in much deeper waters; both of these will pose enormous problems for MCM forces. Other likely developments include increasing the mine's effectiveness against specific targets, in particular minehunters, and seeking to capitalise on influences so far not developed. For example, some mines are now being programmed to react to a minehunter's sonar, or for the tethering cables of moored mines to react to the operations of an ROV and then to set off a chain reaction of other mines in an attempt to destroy the minehunter.

On the MCM side, more efficient mine detecting sonars and command and control systems will be developed to deal with the mass of data accumulated. Efforts will be made to reduce reliance on human operators, for minehunting can be an extremely tedious task and, when operators become bored or tired, fatal errors can be made. Greater efforts will be made to make MCMVs more immune to mines, and new generations of submersibles will be developed with much greater capacities for dealing with more sophisticated mines, longer mission times, increased speed and range and deeper diving capabilities. Also, smaller ROVs will become attractive as complementary systems to the larger vehicles.

The use of datalinks to relay information between vessels on task and the shore–based MCM headquarters is now becoming essential to the speeding up of MCM operations and ensuring that shipping is allowed to use cleared channels at the earliest opportunity. Sonar prediction techniques, already widely used in ASW operations, will also become more important as a means of improving effectiveness and safety with the ability to define safety circles more carefully. Finally, the requirement to keep the MCM platform as far removed as possible from danger zones, particularly in areas which have not previously been route surveyed, may well lead to the further development of remote–controlled, autonomous surface and subsurface platforms for minehunting and disposal.

This report examines in detail the current status of MCM forces around the world. Regional forces are examined country by country, the current status of the force being described, together with known future plans and forecasts of possible upgrading and acquisition of replacement or new hulls. This survey is backed up by tables listing MCMV strengths in 10 year cycles since 1946, indicating how forces have grown and if they are currently growing or remaining static. These are also analysed by design.

This is followed by a detailed study of current designs on offer, together with an assessment of their current market prospects. Tables enumerate main specifications and equipment and major export sales are also listed.

These two main studies are followed by chapters covering the main equipment areas of the MCMV – combat systems, sonars, underwater mine warfare vehicles, shipborne minesweeping systems and machinery. Systems are briefly described together with tables listing where systems are in service and the number of units sold. Market prospects are also examined.

The report concludes with an examination, region by region, of known, current and future programmes of construction and an assessment of possible future requirements based on the age of current operational units.

Bad Bevensen, one of the German Navy's coastal minehunters of the Frankenthal class *(H M Steele)*

The Netherlands Alkmaar class minehunter Middleburg *(H M Steele)*

Myosotis, a Belgian Navy Flower class coastal minehunter *(H M Steele)*

The Saudi Arabian minehunter Al Jawf *(H M Steele)*

NAVAL MINE WARFARE MARKETS

SECTION 1 - BACKGROUND

1.1 Post Second World War Mining Campaigns

Writing in the February 1991 edition of Military Technology the late Desmond Wettern noted that *'One weapon has featured more frequently and consistently than any other in conflicts at sea since 1945: the mine. Yet, to judge by the response of many navies to the threat it still remains a secondary issue demanding far fewer resources than, say, anti-submarine warfare or, indeed, submarines and their systems.'*

It is a well known fact that the mine is probably one of the most cost-effective weapons in the naval arsenal. It is small, relatively cheap (compared to other weapon systems), easy to hide, easy to store and can be laid from virtually any type of platform - ship or aircraft, of any type. It is also the most difficult of weapons to counter. The environmental conditions inherent upon its deployment necessitate some of the most sophisticated technological equipment that man can devise to counter it. The mine is analogous to the 'Fleet in Being' i.e. 'A Weapon in Being' or the 'Weapon that Waits'.

Because of the factors inherent both in its physical form and in its psychological use, the mine is the perfect clandestine weapon to be used either for defensive or offensive purposes, either through 'blackmail' (i.e. declaring an area to be mined when it is not) or by posing an actual threat. The deployment of mines requires a counter effort to neutralise them that is out of all proportion to the effort put in by the force which laid the weapon. It is also true that the mine creates a psychological effect out of all proportion to its cost and size.

History abounds with stories about mine warfare, its effects, the potency of the system, its psychological effects and so on.

A close study of past mining campaigns only serves to emphasise that, other than technological advances in the weapon and the countermeasures to it, previous experience of mine warfare is as valid today as when the weapon first began to be used.

And yet, because of the nature of the weapon, and the unglamorous nature of the countermeasures, MCM has, in the past, been placed lower down on the list of priorities of many nations naval requirements.

Set against the background of history it seems inconceivable that navies should treat the mine threat with such a cavalier attitude as they have. It is, perhaps, just worth briefly noting some of the statistics that have highlighted the influence of the mine this century.

During the First World War, 235,000 mines were laid in European waters of which Germany laid 50,000. The North Sea Barrage laid in 1917-18 extended some 240 miles from Scotland to Norway. Consisting of 70,000 mines this field closed the German Navy's exit to the North Atlantic. Altogether, mines accounted for damage or loss to some 685 ships of over 1 million tons, including 102 German warships.

During the Second World War a total of 635,000 mines were laid resulting in the loss of 1.4 million tons of merchant shipping. British, allied and neutral shipping losses caused by mines amounted to more than 1.1 million tons. Altogether a total of 5,906 ships were lost or damaged through mines.

TABLE 1.1
Analysis of Cause of Loss of Shipping Tonnage in the Second World War

Cause	Percentage
Mine	35%
Submarine	40%
Surface Ship	16%
Aircraft	7%
Other	2%

In the Far East the mining of Japan's 'inner' and 'outer' seaward defence zones with 12,000 mines under 'Operation Starvation', which commenced in April 1944, resulted in the sinking or severe damage of more than 650 ships, including 65 warships. By this stage of the war Japan relied heavily on sea imports (oil 80%, iron 88%, and food 20%) in order to maintain its ability to conduct war, and sustain the population. 'Operation Starvation' virtually crippled that ability. The US Air Force estimated that a reduction of 20–30% in the import of foodstuffs would be the difference between actual subsistence and starvation for a large part of the population. In late 1945 a report from the Japanese civilian authorities to the military warned that if the war continued for another year, 7,000,000 Japanese might be expected to die of starvation.

Throughout the war in the Far East the US laid nearly 31,000 mines which led to the destruction of 1,075 Japanese ships with a total tonnage of 2,289,000 tons. This mining campaign had such an impact on Japan that a former Japanese navy commander, Commander Saburo Tadenuma, claimed that it was *'one of the main causes of our defeat'*.

During the 1950s and 1960s, mines laid during the First and Second World Wars were regularly washed up on North Sea beaches in the wake of North Westerly storms. The North Sea was not formally declared free of mines until the late 1960s, more than 20 years after the end of the war.

During the Korean War in 1950–52, North Korean junks and sampans laid 4,000 Soviet antiquated mines in the approaches to Wonsan harbour, blocking a US/UN invasion fleet of 250 ships for eight days. US minesweepers, together with South Korean and eight Japanese minesweepers, were called in to clear the mines. Two US and one South Korean minesweeper were sunk. As a result the amphibious landing was delayed for almost a week, because the nine days allowed for minesweeping proved insufficient. When the landing was finally made, ground forces had already taken Wonsan.

Rear Admiral Allan Smith, commander of the amphibious task force summed up his frustration in a report to the Chief of Naval Operations:
"We have lost control of the seas to a nation without a navy, using pre–the First World War weapons, laid by vessels that were utilised at the time of the birth of Christ." {Tamara M. Melia – *'Damn the Torpedoes': A Short History of US Naval Mine Countermeasures, 1777–1991*, Washington DC, Naval Historical Center, 1991}.

Because of the inability of the surface vessels to effectively sweep the mines, the Americans decided to consider the helicopter as a possibly more cost–effective alternative.

Following their experience of mine warfare in the Korean War, the military during the Vietnam War fully endorsed the importance of mine warfare. Admiral Thomas H. Moorer, former chief of naval operations and chairman of the Joint Chiefs of Staff (1970–74) stated:
"I tried over and over again [during the Vietnam War] to get authority from Washington to mine Haiphong harbour."

However, it was not until early May 1972 that President Richard M. Nixon gave the Navy permission to carry out a mining offensive. At that time Navy, Marine and Air Force pilots were flying a combined total of 1,000 sorties a day over Vietnam in an effort to prevent the Viet Cong from supplying their forces at the battlefront. Noting this, Admiral Moorer said:
"On this particular morning, [the first mining sortie] it took only one aircraft carrier and 26 aircraft to mine Haiphong harbour. It took them less than one hour.
"Not only was no one injured, not one ship entered or left Haiphong after the mines [36 in total] were dropped until our own US minesweeping forces removed the mines from the channel after the war."

Three major ports (Haiphong, Hon Gay and Cam Phon) were mined by US Navy aircraft, trapping 50 ships. Because the North Vietnamese lacked the necessary assets to clear the mines, the harbours remained closed to shipping until the Americans swept the mines after the war, some 300 days later. The mining cut North Vietnamese imports by 30% and reduced supplies reaching the combat areas by an estimated 800 to 1,500 tons per day. The after effect of this small operation forced the Communists to the negotiating table.

During the Yom Kippur War in October 1973, it was claimed that mines had been laid in the Suez Canal. After the war, Britain, France and the USA sent MCM forces to carry out minesweeping operations in the Canal. Not until 1975 was the 70 miles of Canal finally declared clear. Vast quantities of explosives were swept from the Canal, but no mines were found. The mere threat of mines had considerably hampered military operations in the area.

The 1980–1988 Gulf War between Iran and Iraq witnessed a rather different type of warfare. Both sides realised that to embark on a mining campaign might possibly do equal harm to their own interests as well as to those of the enemy. Hence, mines did not play a large part in what became known as the 'tanker war'. Instead, the protagonists were content to rely on missiles which, while subjecting the enemy's maritime trade to harassment, did not, in effect, hinder the maritime trade of the attacker. A mining campaign might well have closed the Gulf to all shipping, thus denying both sides the ability to carry on a maritime trade, on which both heavily depended for their existence and valuable foreign income.

Nevertheless, mines were employed in the Gulf during the war, and the first reported mine casualty occurred on January 11, 1982. Not until July 1987, however, was a major casualty claimed, a US ULCC. During this year US–flagged tankers began to be escorted by a US naval task group which was supported in its operations by a US force of minesweeping helicopters. During these operations the US frigate *Samuel B. Roberts* was damaged by mine in 1988.

In 1982 British MCM forces were deployed to the South Atlantic in support of naval operations during the South Atlantic War with Argentina. It was in the wake of this conflict and its use of four deep–sea trawlers taken up from trade for use as MCM vessels to sweep for possible mines in the waters in and around Port Stanley and other areas, that the COOP (Craft of Opportunity) concept came into being. The experience led the Royal Navy to return to the minesweeper, and a class of deep–sweep vessels was commissioned into service.

The most recent conflict to involve the use of sea mines was Operation Desert Storm, the Coalition War with Iraq in 1991. Naval forces played a major part in this war, and it soon became apparent that there was a real threat from Iraqi–laid sea mines. Some 20 mines were sighted in the Gulf on January 16 and immediately minesweeping operations commenced. Forces from Australia, Britain, Canada, France, Saudi Arabia and the USA took part. Sadly they were unable to prevent two major US warships from suffering severe damage from mines – the amphibious landing vessel USS *Tripoli* and the cruiser USS *Princeton*. Once again, the mine potentially caused a major turnaround in maritime strategy, for a possible amphibious landing off Kuwait was definitely called off soon after the mining incident. Whether the mines directly affected the decision to abandon the amphibious landing, or whether this was a result of other tactical considerations, has never been revealed. One certain effect that the Iraqi mines had, however, was to interfere with US and Coalition forces maritime control of the northern Persian Gulf. The mines hampered the sealift, including the transport of cargo bound for Saudi ports, and also prevented the battleships *Wisconsin* and *Missouri* from manoeuvring freely to provide gunfire support to forces ashore until the channels were swept.

1.2 Effects Mining Campaigns After 1945

How, then, have the operations briefly noted above affected the development and raison d'être of mine countermeasures forces since 1945? The end of the Second World War saw the decommissioning of large numbers of warships, although large numbers of mine warfare vessels were retained in Europe for a while to clear European shipping channels of known minefields.

TABLE 1.2 Major Mine Warfare Forces Order of Battle 1946–1996

Country	1946	1956	1966	1976	1986	1996
EUROPE & SCANDINAVIA						
Albania	3	3	22	18	15	3
Belgium	0	37	51	29	22	9
Bulgaria	0	23	31	20	33	20
Denmark	64	30	26	10	6	9
Finland	48	48	6	9	13	13
France	31	126	99	60	36	22
Germany – East	0	48	22	55	28	–
Germany	0	6	99	61	60	58
Greece	31	31	27	14	14	19
Italy	11	77	60	46	29	12
Netherlands	25	53	68	38	25	17
Norway	11	13	10	10	8	7
Poland	13	23	53	44	32	22
Portugal	0	16	16	4	4	0
Romania	0	4	36	24	47	43
Soviet Union/Russia	115	25	369	372	376	173
Spain	7	19	25	22	12	12
Sweden	19	25	23	38	38	32
Turkey	2	8	12	25	26	29
UK	435	176	115	44	45	23
Yugoslavia	12	30	31	18	14	15
TOTAL	**827**	**821**	**1201**	**961**	**883**	**538**
MIDDLE EAST, NORTH AFRICA, GULF & WESTERN ASIA						
Egypt	3	14	16	12	6	15
India	11	12	10	8	16	18
Pakistan	0	5	8	7	3	9
TOTAL	**14**	**31**	**34**	**27**	**25**	**32**
ASIA, PACIFIC & AUSTRALASIA						
Australia	32	19	6	3	1	10
China	0	4	23	25	148?	95
Indonesia	0	18	23	7	2	13
Japan	0	35	34	26	37	35
Korea (North)	0	0	0	0	0	23
Korea (South)	0	17	14	13	13	14
Taiwan	0	8	8	14	13	13
Vietnam	0	3	3	2	0	15
TOTAL	**32**	**104**	**111**	**90**	**214**	**218**
NORTH AMERICA						
Canada	38	44	6	6	6	12
USA	148	185	128	38	22	22
TOTAL	**186**	**229**	**134**	**44**	**28**	**34**
GRAND TOTAL	**1059**	**984**	**1480**	**1133**	**1150**	**822**

The Korean War experience, when MCM forces were found to be totally inadequate to meet the real or intended threat, led to a major re–appraisal of MCM forces in many navies. The result of this can be seen in Table 1.2 above.

Although large numbers of ships, including mine warfare vessels, were being decommissioned and scrapped in the years following the ending of the Second World War, the impact of the mining campaign during the Korean War led to the construction of a whole new generation of MCMs built in the mid–1950s and 60s. These were the British 'Ton', the Soviet T 43 and the American 'Adjutant' and 'Aggressive' classes. Apart from the large numbers built for domestic use, many were also supplied to friendly navies.

The effect of other mining campaigns noted above can also be seen in the response by various navies to the acquisition of mine warfare vessels in the immediate aftermath of a particular conflict. This response, however, has not always been of the positive kind. Many navies continue to ignore the lessons of these campaigns and almost resolutely refuse to encounter the acquisition of mine warfare vessels. Sometimes this has been to their cost. Others have read the wrong lessons into the situation. The USA, for example, in the aftermath of the Korean War, considered that surface mine warfare forces were not effective in the face of an intensive mining campaign, even if the mines were of a very simple horned type. This, together with the increased use of mines operating on acoustic and magnetic influences led the Americans to decide that the helicopter might offer a better and safer platform from which to conduct MCM operations. Furthermore it would probably cover the ground at a faster rate, necessary if the minefield was an extensive one with large numbers of mines in it. This method of conducting MCM operations has not always proved to be as effective as had been hoped.

In other cases nations have tended to rely on those navies with considerable MCM expertise to carry out any mine countermeasures operations on their behalf.

In the final analysis it has to be said that there is no substitute for possessing ones own MCM assets, and in peacetime carrying out a vigorous plan of route surveying (see below) to ensure that one is continually aware of the state of the shipping channels and the seabed below with regard to any possible mine threat. Such foresight will go a long way to ensuring that a nation is not caught unawares by a mining campaign, and if one is carried out, then the assets are readily available to deal with that threat before it reaches the state where the nation is forced into an uncompromising position.

1.3 Mine Countermeasures

Modern mines are far removed from the Second World War common moored contact type, although there are still large numbers of this type around which are just as deadly as was proved in the Gulf. The most common types today are influence mines (both ground and tethered) which are designed to react to either magnetic, acoustic or pressure signatures or a combination of these.

The key to mine warfare is the ability to define the threat. The more that is known about the threat, the easier it is to counter. There is little that is not already known about the present day threat covering the moored mine, various types of ground influence mine and the volume threat posed by devices such as rising mines and so on. The problem is actually finding and dealing with these sophisticated weapons

Mine countermeasures are principally carried out by specialist mine countermeasure vessels (MCMVs) equipped with mine detection sonars, minesweeping equipment and/or remote operating or autonomous vehicles which can detect mines, lay charges against them and detonate them from a distance. Divers are also employed in mine disposal operations.

Mine countermeasures techniques fall into two main areas: active and passive.

Active MCM

Active MCM is divided into two main spheres of operation: minehunting and minesweeping.

Minehunting has proved to be the only relatively safe and effective method of dealing with modern sophisticated influence mines. If the minehunter is to be effective in its task however, it must be equipped with highly sophisticated equipment. In order to be able to pinpoint the exact location of a mine and to record its position precisely requires accurate navigation systems of the highest order. Next, in order to detect the mine and carry out accurate classification, the vessel must be equipped with an efficient sonar system able to detect the smallest of targets under the most adverse conditions. Having detected and classified contacts, the next task is to neutralise those classified as mines. The main task of mine destruction is now carried out by small submersibles – Remotely Operated Vehicles (ROVs) or Autonomous Underwater Vehicles (AUVs) deployed from the minehunter. These are either used to sweep ahead of the vessel to detect mines, or to carry out precise classification of objects detected by the minehunter and then to position a remotely detonated countermining charge next to the mine. Alternatively they may be equipped with powerful cutters which sever the tethering wire of moored contact mines so that they float to the surface where they can be destroyed. However, this method of dealing with moored mines does not always prove successful, and is extremely time consuming. This is an instance where clearance divers can be used with great effect, as long as the mines are not fitted with sophisticated anti–personnel devices.

The second active MCM method is minesweeping. This involves systematically sweeping a defined area to destroy any mines which may be in it without the need to identify them. Minesweeping involves the use of towed wire sweeps to cut the wires of moored contact mines so that they float to the surface where they can be destroyed. Influence sweeps, in which noise makers and large cables carrying electric current to create magnetic fields are towed behind the minesweeper still remain effective against the less sophisticated ground mine. Other methods include sweeps towed by helicopters and remote controlled sweeps such as the German Troika and Swedish Sam systems. In these systems small craft carrying noise generators and powerful electric motors to create magnetic fields proceed ahead of the control craft, which is usually itself an MCMV. The deep sweep whose operating depth can be adjusted during the sweep operation has been developed to counter mines laid in deep water. To carry out this operation the minesweepers have to operate at least in pairs. This method of mine countermeasures requires extremely accurate navigation and station–keeping, and such vessels must be equipped with the most up–to–date navigational aids available if they are to perform their task effectively.

Minesweeping, therefore, still plays a very important part in any active MCM operation. Vast numbers of moored mines of all types, both contact and influence, are still held in inventories around the world. In consequence ever more sophisticated means of minesweeping need to be deployed in order to be able to counter any possibility that the latest types of these mines may be used against a nation.

The key to effective MCM operations (both minehunting and minesweeping) is accurate navigation and precise positioning. It is often necessary for vessels to subsequently return to the position of previously defined contacts to re–examine them or, in the case of confirmed mines, to carry out countermining. Precise positioning and plotting are absolutely essential if an MCMV is to carry out this task effectively. With such a system a vessel can return to the precise position where a contact was previously found with considerable saving in time and with the sure knowledge that the original contact will be found again.

Passive MCM

The most effective passive form of MCM is that of route surveying. In peacetime this is a primary task not only of the MCM force but also of the hydrographic service. To be effective, route surveying requires that the whole of a proposed wartime shipping route be surveyed extremely carefully to produce bottom contour charts which show in the minutest detail the composition of the seabed and precise data on all objects on it. By regularly surveying and updating the charts, accurate pictures of proposed routes can be maintained which will offer safe passage for those selected for use in wartime. Furthermore, these routes can be checked by the minehunters much more rapidly and accurately as the vessel then only has to check objects not already marked on the chart.

Other forms of passive MCM include such techniques as noise reduction (reducing cavitation to a minimum, reducing machinery noise and so on), deperming and degaussing (to reduce a ship's permanent and induced magnetic signature to a minimum) and optimising the hydrodynamic shape of the hull with a minimum displacement (to reduce pressure signature to a minimum). Alternatively, it may be possible to modify a ship's signature so that it no longer appears to be what it actually is. Such techniques, however, require that one can precisely define the threat. Mine avoidance is another passive technique which can be employed, but again requires prior knowledge of the precise nature of the threat.

HMS Hurworth of the UK Royal Navy Hunt class of minesweepers/minehunters *(H M Steele)*

HMS Ledbury, the second of the UK Royal Navy's Hunt class *(Crown Copyright)*

SECTION 2 - MINE WARFARE FORCES - FLEET STRENGTHS

An examination of Table 2.1 below clearly shows the current state of mine warfare forces world-wide on a regional basis.

TABLE 2.1 World Inventory Mine Warfare Vessel Strengths*

Region	Region 1	Region 2	Region 3	Region 4	Region 5	TOTAL
Units	597[§]	83	240	66	18	**999**

* Includes COOP (where known), river clearance vessels, drones, route survey vessels, MCM support vessels and tenders, but does not include minelaying vessels
§ Includes six MCM modules allocated for Danish '*Flyvefisken*' class
Region 1: Europe & Scandinavia
Region 2: Middle East, North Africa, Gulf & Western Asia
Region 3: Asia, Pacific & Australasia
Region 4: North America, South America & Caribbean
Region 5: Africa

2.1 The Current World Market

As noted in the previous chapter, improving mine warfare forces capability usually takes place in the aftermath of a mining campaign, rather than before. For a few years intensive effort is put into building up mine warfare forces following a mine threat, with additional assets being acquired. Then, as the threat passes into history, MCM forces once more fall into decline.

Not surprisingly the region with the greatest number of MCM assets is Europe and Scandinavia. The effects of two major mining campaigns this century in European waters have done much to heighten the awareness of the threat.

Germany, Russia and the UK in particular have first hand experience of what both offensive and defensive mining campaigns can accomplish. Hence, these three nations have paid close attention to the mine warfare scenario since 1945.

Since 1945, however, the UK too has seen a major decline in its MCM assets. During the 1970s and 80s Cold War era, the threat of a mining campaign to the economic independence of the nation was tempered by the need to ensure the safety and ability of the Navy to deploy the nation's nuclear deterrent. Hence, much of the priority in the mine warfare effort was targeted at ensuring that the exit routes and bases for the strategic nuclear submarine fleet were free from the threat of mines, rather than ensuring the maximum number of safe routes for mercantile trade. In spite of this shift in emphasis, however, considerable effort was put into planning and providing for a select number of mine-free channels into UK ports for merchant shipping.

Thus, while most navies, and in particular the North European navies of NATO and the navies of the Warsaw Pact, maintained strong MCM forces throughout the height of the Cold War, since the beginning of this decade there has been a steady decline in the number of vessels dedicated to MCM. This has been particularly noticeable among the former Warsaw Pact navies and especially in Russia, where the MCM fleet has more than halved in the last 10 years, and much of what remains is of obsolescent design.

However, numbers alone do not give an accurate indication of the ability of a navy to ensure mine-free shipping channels in time of rising tension and war. With today's sophisticated technology supporting the MCM fleet, and the ongoing peacetime task of route survey closely supported by a well-organised and well-equipped hydrographic service, much of the impact of a sudden unannounced mining campaign can be neutralised in advance.

Where such a threat – real or not – becomes uncontrollable is when the assets to hand are insufficient to ensure a regular uninterrupted flow of merchant traffic into various ports, together with the ability of the navy to move freely in and out of its bases. For example, some 400 merchant ships a day dock in UK ports alone. For each navy to ensure such a situation does not become untenable demands a finite number of assets and there must be a question mark over the scale of some European navies assets. Such a dilemma – taking into account the financial climate – does not even address the question of the technological superiority of such forces.

For the countries of Europe and Scandinavia the conundrum remains – will they ever again face the threat of a major mining campaign in European waters? Obviously there are some nations who have concluded that there is no such possibility, while others are doing their best in the present climate of international rapprochement to ensure that they are not caught completely unprepared again.

Denmark, Germany, Norway, Spain and the UK are currently in the process of ensuring that, while overall numbers may remain static, the qualitative capability of their MCM forces remains unsurpassed. New vessels are replacing obsolete units, while existing units that are not overage are being upgraded with the latest technology.

The future prospects for MCM forces in this region are mixed. Much of the willingness of some navies to upgrade their MCM forces will depend on the financial climate of the nation. This in turn will effect how the navy itself views its commitments, priorities and in turn its overall quantitative and qualitative assets. Mine warfare forces are not glamorous, and to aspiring captains of warships, command of an MCM unit, squadron and even indeed in some cases command of the entire MCM force, is sometimes seen as a bit of a dead–end career. Many see rostering in the MCM force merely as a step on the path to greater things, command of a large warship or submarine, for example.

Scandinavian navies are especially concerned about the state of their MCM forces, for these countries are particularly vulnerable to a mining campaign, relying as they do on mercantile trade for the main source of their revenue.

The Mediterranean countries too, have accepted the need for modern high–tec MCM forces. Italy especially has developed a highly modern MCM force, and with its MCMV designs, has captured a major share of the export market.

Portugal remains the one European navy where adequate MCM assets are sadly lacking. Financial problems are primarily responsible for this state of affairs – but the situation will need to be addressed before the end of the decade.

After Europe and Scandinavia the region with the next largest MCM forces is Asia, the Pacific and Australasia. This region has, since 1945, witnessed a steady increase in MCM strength, with a number of navies accepting the importance of having a well equipped MCM fleet.
Not surprisingly in view of its experience during the Second World War, Japan, since forming the Maritime Self Defence Force after the war, has maintained its MCM fleet at an almost constant level of about 35 units, consistently replacing obsolete units with new construction.

The navy with the largest number of MCM assets in the region is China. But the overall strength of China's MCM fleet has declined slightly over the last 10 years.

Following its experience during the Korean War of 1950–52, South Korea too has maintained a MCM fleet of around 14 units During the last decade obsolete units have been replaced with new domestically built vessels equipped with the latest technology.

Other navies too are developing and modernising their MCM fleets to keep abreast of the latest technology to ensure that they will be able to counter the most modern, technologically sophisticated mines which they consider they might encounter. Malaysia, Singapore and Australia have either recently or are in the process of modernising and upgrading their MCM fleet.

With the future stability of this region uncertain it is apparent that the major navies will wish to maintain their current MCM assets at a credible level and in a proper qualitative state technologically. As navies begin to acquire a much greater 'blue water' capability with more high value modern units such as frigates, destroyers, submarines and even aircraft carriers, it will become ever more important that they possess suitable and adequate MCM assets purely to protect those important warships, as well as maintain the ability to ensure mine–free approaches to their major ports and naval bases.

This will certainly be the case bearing in mind the huge arsenal of mines held in the inventories of China and North Korea – even if large numbers of these mines are of the obsolete moored contact variety.

The region with the next largest MCM forces is the Middle East, North Africa, the Gulf and Western Asia. This region is one where there are very mixed views concerning mine warfare. Recent history has shown that the restricted waters of the Gulf, Red Sea and Suez Canal, all of them major waterways conveying an enormous amount of merchant shipping, are particularly vulnerable to a mine threat. In the face of such a threat it is interesting to note that only two Middle Eastern states bordering these waterways have paid any attention to their MCM capability – namely Egypt and Saudi Arabia, both of whom are developing a modern MCM capability. In the Gulf the majority of navies seem to prefer to leave any MCM effort required to friendly navies, predominantly western, to conduct on their behalf. To meet any such contingency the US has stationed two minehunters permanently in the Gulf.

Iran, a country with a massive stock of mines which can be laid from any number of different platforms, possesses a very meagre MCM capability.

Round in the Indian Ocean, with continuing tension between India and Pakistan, both countries have built up capable MCM forces and will maintain these in a high state of readiness to meet nay possible mine threat.

The only other nations with any pretence to deploying MCM forces are Syria and Libya. The state of the latter is very much open to question, while the former is probably in dire need of modernisation if it is to retain any form of credibility against a mine threat. However, it is unlikely that Syria will face a major mine threat as it does not rely heavily on mercantile trade for economic viability, and so the need for a large MCM force does not at present exist.

As far as the North African states are concerned they have, apart from Libya, paid virtually no attention at all to any potential mine threat. Sea mines might well form a part of the terrorists arsenal in the future, and Algeria, Morocco and Tunisia will probably see increasing tension in their countries as the threat from Islamic fundamentalists grows. In the face of any such hostility to the current regimes, the ports and naval bases of these countries will be particularly vulnerable to a mine threat and their navies would be well advised to give careful consideration to building up a modern MCM capability.

The South American, North American and Caribbean region also shows very mixed views towards MCM. The South American navies have, on the whole, paid little or no attention to the need for effective MCM forces. The one nation which has is Argentina. Having built up a sizeable mine arsenal, a proportion of which was used in the South Atlantic War of 1982, the Navy also developed an MCM capability. Since 1982 and the subsequently dire financial state of the economy, this capability has not kept in step with advancing technology. The Argentine Navy undoubtedly recognises the need for an adequate MCM capability and any spare resources ought to be devoted to upgrading this capability, which can also be of value when used in a patrol configuration, as is the case with many other navies' MCM forces.

In view of Argentina's recognition of the important role that mine warfare can play in a maritime strategy, it is surprising that other South American navies have not followed its lead. In most other areas of naval warfare the major navies of South America have kept very much on a par with each other. Only Brazil and most recently Uruguay have given heed to the need for an MCM capability. In neither case, however, can the assets be considered to be modern by today's standards.

All in all, such MCM assets that the South American navies possess must be considered obsolete by modern standards and sorely in need of upgrading if not immediate replacement. Other navies in the region would be well advised to consider acquiring an MCM capability.

In North America the USA has never really faced a major mine threat in national waters. The sum total of America's mine warfare experience has been in overseas operations. From a mine countermeasures point of view much of the US Navy's expertise developed out of its involvement in the Korean War and Vietnam. Because of those experiences much of the MCM effort was devoted towards developing airborne countermeasures, with the surface MCM fleet declining rapidly since 1966. The current strength rests at 22 units, which comprise two modern classes of MCMV. Following some initial problems these are now proving to be much more effective in their assigned role. Construction of the latest class is continuing. Meanwhile, considerable effort is now being devoted to developing MCM for deployment in littoral areas in support of amphibious operations. These developments have been spurred on by the American's experience in Operation Desert Storm, where mine damage to an amphibious vessel may have played a major part in the decision to call off a possible amphibious landing in Kuwait in 1991. Apart from various countermeasures developments, the Americans are also considering the deployment of air cushion vehicles in the MCM role. To gain experience and build up a pool of MCM expertise, the US is rotating crews to the two minehunters stationed in the Gulf (see above).

Canada too is building up a modern force of MCMVs and may well be the first navy to deploy the AUV concept, using a drone craft ahead of the MCMV to direct the operations of a small submersible.

The last region to be covered is Africa. Here only two nations can be said to have paid any real heed to the need for MCM forces – South Africa and Nigeria. In the latter case, there certainly is a need for MCM in view of the major natural oil resources on which the country's economy largely depends. However, there is always a suspicion that the acquisition of the modern Italian-built MCMVs was more for prestige than any desire to meet a real need. Like much of Nigeria's other naval assets, these are now, sadly, greatly in need of maintenance in order to provide an effective MCM capability.

South Africa has managed, in spite of the arms embargo, to maintain an MCM fleet in a high degree of readiness and modernity. However, there is a real need to replace the aged hulls with new construction, and there will doubtless be moves in this direction by the end of the decade.

As for the rest of Africa, the other navies are so poor, and in spite of hostility between a number of nations, there is very little real threat of any mining campaign between states. The real threat to many of these states comes from land rather than naval operations, and so the need for MCM forces is very minimal.

All in all there is considerable need around the world for many navies to either build replacement MCMVs, upgrade existing units, or expand existing forces with new construction.

That MCM should rate high on any navy's list of priorities should not be in question. The previous chapter has outlined how vulnerable any nation is to a mining campaign, especially if MCM assets are inadequate to meet the threat. Figures prepared by the UN show that world maritime trade in 1994 reached a record 4.46 billion tons, an increase over 1993 of 3%. It is anticipated that 1995 will show a similar increase of 3%. To cope with the growth in maritime trade, the overall size of the world's merchant fleet increased by a factor of 3:2. With these figures likely to continue increasing, and with the proliferation of mines and mine types in naval inventories around the world, it is essential that adequate and up-to-date MCM assets are available to any maritime nation.

2.2 Regional Forces 1945–96

2.2.1 Europe & Scandinavia

ALBANIA

The two ex-Soviet T 43 type minesweepers transferred in the 1960s are obsolete and probably non-operational. As Albania no longer enjoys the support of either Russia or China maintaining these craft may be beyond the navy's capability. It is unlikely that any new MCM craft will be acquired in the near future, unless funds are made available.

BELGIUM

In 1989 Belgium signed a MoU with the Netherlands to build a new generation of GRP-hulled minesweepers. Portugal joined the project in 1991, but in 1993 the Netherlands withdrew, while Portugal with limited funds available, decided to proceed with other priorities. Belgium continued with the project on its own and authorisation for the acquisition of four units was given in 1994, work on the first being scheduled to commence in 1997 with completion due in 1999. The ships will be equipped with sophisticated minesweeping equipment including the French Thomson Sintra ASM Sterne M magnetic sweep equipment, evaluation of which began in 1995. The main strength of the current MCM force consists of seven *'Tripartite'* minehunters developed in co-operation with France and the Netherlands, France supplying the MCM equipment and electronics, the Netherlands the engine room equipment and Belgium the electrical installation. Three of this class are being upgraded with improved sonar and ROV. Two of the old

'Aggressive' class ocean minesweepers are being kept operational for the time being, although they are well beyond their life expectancy.

BULGARIA
A number of ex–Soviet minesweepers are operational, but all are obsolete and need replacing. The last two PO2 class inshore minesweepers acquired in the 1950s and 60s will shortly be paid off. Other obsolete vessels being paid off include the inshore minesweepers of the *'Yevgenya'* type transferred in 1977, and the *'Vanya'* class coastal minesweepers. The latest types to enter service are the *'Sonya'* class transferred from the USSR between 1981–84 and five *'Olya'* class built in Bulgaria between 1988 and 1992. The sixth unit of this class was completed in 1995. Further units may be built to replace obsolete vessels now being paid off. A force of about 20 minehunters/sweepers is planned with an in–service date around the year 2000.

DENMARK
A modern MCM capability is being developed using the Stanflex 300 *'Flyvefisken'* class multi–role design and the SAV drone craft. The Stanflex concept uses 'plug in' modules so that the force can be rapidly expanded or reconfigured for other roles. Six MCM modules are available. Two SAV drones are operational and four more were ordered in mid–1994 for delivery in 1996/97. A final batch of four is planned to enter service in 1988, making a force of 10 craft. The drones deploy a towfish with side scan sonar. The obsolete *'Bluebird'* class coastal minesweeper is used in the surveying role. Near term future requirements are likely to be constrained to upgrading the *'Flyvefisken'* class MCM capability which can be easily and simply achieved using the Stanflex module concept. There is not likely to be any requirement for new hulls at least until towards the end of the first decade of the next century.

FINLAND
There is a small MCM force of 13 inshore minesweepers. The *'Kuha'* class, completed between 1974 and 1975 are to be equipped with a new mine sweep system developed by Fiskars. The *'Kiiski'* class were all completed in 1984. Plans to replace these craft will probably be drawn up around the middle to end of the next decade.

FRANCE
The bulk of the MCM force is allocated to protect the SSN and SSBN force, but overall MCM assets are insufficient to provide effective cover for all bases and shipping routes. A major construction programme to modernise the MCM force was cancelled in 1992, and the five *'Circe'* class were retained in service. However, these are now to be withdrawn from service in 1997–98. No decision has been taken regarding their replacement, and under the present stringent financial climate it is unlikely that any such decision will be taken in the near future. Ten *'Tripartite'* minehunters built in collaboration with Belgium and the Netherlands are operational, and an eleventh unit is under construction to replace a unit transferred to Pakistan in 1992. Two other units are on order for Pakistan. A programme to upgrade the sonars, navigation systems and ROVs of the French *'Tripartite'* vessels to bring them up to the same standard as their sister ships in the Belgian and Netherlands navies was begun in 1995. In addition France has in service three *'Antares'* class vessels used for route surveying and four diver support tenders.

TABLE 2.2A Mine Warfare Vessel Order of Battle 1946–96§ EUROPE & SCANDINAVIA

Country	1946	1956	1966	1976	1986	1996
Albania	3	3	22	18	15	3
Belgium	0	37	51	29	22	9
Bulgaria	0	23	31	20	33	20
Denmark	64	30	26	10	6	11
Finland	48	48	6	9	13	13
France	31	126	99	60	36	26
Germany (East)	0	48	22	55	28	–
Germany	0	6	99	61	60	58
Greece	31	31	27	14	14	19
Hungary	0	0	0	0	0	51
Italy	11	77	60	46	29	12
Latvia	0	0	0	0	0	2
Netherlands	25	53	68	38	25	17
Norway	11	13	10	10	8	7
Poland	13	23	53	44	32	22
Portugal	0	16	16	4	4	0
Romania	0	4	36	24	47	43
Soviet Union/Russia[1]	115	25	369	372	376	173
Spain	7	19	25	22	12	12
Sweden	19	25	23	38	38	32
Turkey	2	8	12	25	26	29
UK	435	176	115	44	45	23
Yugoslavia	12	30	31	18	14	15
TOTAL	**827**	**821**	**1201**	**961**	**883**	**597**

§ Figures in all tables are projected to the end of 1996. Figures include COOP (where known), river clearance vessels, drones, route survey vessels, MCM support vessels and tenders, but does not include minelaying vessels
1 The figures for units listed in the Order of Battle are open to question, but are considered to be representative. They do not include transfers overseas.

GERMANY

Germany possesses one of the largest, most powerful and modern MCM forces in the world. A major programme of new construction to replace units built in the 1950s and 60s is nearing completion. Ten Type 332 *'Frankenthal'* class coastal minehunters are operational and a further two units were ordered in late 1995. The Type 343 *'Hameln'* class coastal minesweepers are to be modernised. Five will be upgraded with a full minehunting capability incorporating the latest MCM sonar and disposable ROVs which are currently being evaluated under the development phase of the MA 2000 programme, an integrated plan for upgrading mine countermeasures equipment. The other five *'Hameln'* class will be converted to include a *'Troika'* drone control capability together with improved mechanical sweeps. These plans are currently in the definition phase. In addition to these very modern units Germany has in service 14 *'Lindau'* class Type 331 and Type 351 coastal minehunters/minesweepers. Six were converted into *'Troika'* drone control ships in 1981/1983. Eighteen *'Troika'* remote control minesweeping drones are operational. The *'Frauenlob'* class Type 394 inshore minesweepers launched in 1965–67 are being paid off, and currently five remain in service. In addition there is an old minesweeper recommissioned last year as a diver support vessel.

GREECE

The MCM force is obsolescent and desperately in need of replacement. The nine ex-US MSC 294 class recently completed a major upgrade with new engines and radar. They may also be given a new sonar in the near future. Acquisition of modern units and equipment is now an urgent priority. Plans exist to acquire a force of three vessels, and *'Tripartite'* minehunters may be acquired from Belgium. The four old *'Adjutant'* class vessels were bolstered by two more of the same class acquired from Italy in 1995. The Greek Navy also operates four minesweeping launches loaned from the USA in 1971 and eventually purchased outright in 1981.

HUNGARY

The Hungarian Navy carries out an important role in patrolling the river Danube. To counter any mine threat to this vital waterway running through Hungarian territory, the Navy operates a force of six *'Nestin'* class river minesweepers built in the former Yugoslavia between 1979–80. In addition the Navy has 45 AN-2 class aluminium-hulled mine warfare vessels which also double as river patrol craft. Completed between 1955 and 1965, these vessels can be transported by road to rivers in Hungary which might be threatened by mines.

ITALY

A major programme of modernisation begun in 1978 is almost complete with the construction of four *'Lerici'* and eight *'Gaeta'* class minehunters. An upgraded *'Gaeta'* design to encompass a multi-role configuration is under consideration and a class of six vessels may be ordered before the end of the decade. Italy has achieved a major export success with this design, selling four to Malaysia and two to Nigeria. A modified design is being built in the USA as the 12 units of the *'Osprey'* class while six more vessels to another modification are under construction for the Australian Navy.

LATVIA

In 1993 Latvia received two Kondor II Type 89.2 class minesweepers of the former German Democratic Republic. Weapons and minesweeping equipment had been removed but there are plans to re-furbish the ships as minesweepers in the fairly near future.

NETHERLANDS

The current force comprises 15 *'Tripartite'* minehunters. Three units of this class are due to be converted to drone control vessels this year, each ship controlling four drones. The *'Troika'* drones will be a modified version of the German system, with the first entering service in 1998. The rest of the class are to have their minehunting capability upgraded with improved sonar and a new mine disposal system. Four of the class will be adapted to deploy propelled variable depth sonars (PVDS). Two additional units were completed for Indonesia in 1988. An MoU with Belgium for the construction of minesweepers (see above under Belgium) was abandoned for financial reasons. Two of the obsolete *'Dokkum'* class coastal minesweepers remain operational.

NORWAY

Construction of nine units under a major new programme is under way to replace the obsolete ex-US minesweepers. Seven units have now been completed and two more are under construction. An option for a tenth unit will not be taken up.

TABLE 2.2B Mine Warfare Vessel Order of Battle 1996
EUROPE

Class/Country	Albania	Belgium	Bulgaria	France	Germany	Greece	Hungary	Italy	Latvia	Netherlands	Poland	Romania	Russia	Spain	Turkey	UK	Yugoslavia	TOTAL
AN–2							45											**45**
M 301																	4	**4**
MCD				4														**4**
MSC 268/294						9									11			**20**
MSR						4												**4**
PO 2			2															**2**
T 43	1												5					**6**
T 301	2											12						**14**
Diver support					1													**1**
Diver support				4														**4**
Tenders															8			**8**
Adjutant						6								8				**14**
Aggressive		2												4				**6**
Antares				3														**3**
Baltika													1					**1**
Circe				5														**5**
Cosar												2						**2**
Cove															4			**4**
Dokkum										2								**2**
Frankenthal					10													**10**
Frauenlob					5													**5**
Gaeta								8										**8**
Goplo											13							**13**
Gorya													2					**2**
Ham																	2	**2**
Hameln					10													**10**
Hunt																13		**13**
Ilyusha													4					**4**
Kondor									2									**4**
Krogulec											3							**3**
Leniwka											2							**2**
Lerici								4										**4**
Lida													22					**22**
Lindau					14													**14**
Mamry											4							**4**

Class/Country	Albania	Belgium	Bulgaria	France	Germany	Greece	Hungary	Italy	Latvia	Netherlands	Poland	Romania	Russia	Spain	Turkey	UK	Yugoslavia	TOTAL
Musca												4						**4**
Natya													27					**27**
Nestin							6										7	**13**
Olya			6										3					**9**
River patrol												25						**25**
River																5		**5**
Sandown																5		**5**
Sonya			4										62					**62**
Tanya													3					**3**
Tripartite		7		10						15								**32**
Troika					18													**18**
Vanya			4										15					**19**
Vegesack															6			**6**
Vukov Klanac																	2	**2**
Yevgenya			4										27					**31**
Yurka													2					**2**
TOTAL	**3**	**9**	**20**	**26**	**58**	**19**	**51**	**12**	**2**	**17**	**22**	**43**	**173**	**12**	**29**	**23**	**15**	**534**

TABLE 2.2C Mine Warfare Vessel Order of Battle 1996
SCANDINAVIA

Class/Country	Denmark	Finland	Norway	Sweden	TOTAL
M15				5	**5**
SAM drone				5	**5**
SAV drone	4				**4**
Bluebird	1				**1**
Ejdern				4	**4**
Gassten				3	**3**
Hisingen				4	**4**
Flyvefisken	6*				**6**
Gilloga				3	**3**
Kuha		6			**6**
Kiiski		7			**7**
Landsort				7	**7**
Oksoy/Alta			7		**7**
Styrso				1	**1**
TOTAL	**11**	**13**	**7**	**32**	**63**

* MCM modules.

POLAND
A mixed fleet of MCMVs exists, with 13 Polish–built *'Notek I'* class minesweepers completed between 1981–91, and three older *'Krogulec'* class. The obsolete *'Krogulec'* units will be removed from service by the end of 1997. Four *'Notec II'* coastal minehunters have recently entered service, but plans to build more of this class appear to have been abandoned. An enlarged design has been developed, but so far no orders have been placed. Two *'Leniwka'* class coastal minesweepers adapted from stern trawlers are also in service.

PORTUGAL
Currently no MCM force exists. Plans to join with Belgium and the Netherlands in developing a new minesweeper (four units were planned) were abandoned in 1993 (see under Belgium above). Once current priorities have been achieved Portugal may rejoin the venture. Development of an MCM force is now an urgent requirement.

ROMANIA
The MCM force was modernised in 1986–8 with four locally built *'Musca'* class minesweepers. Twelve T 301 class minesweepers are also operational. A force of 25 river minesweepers, built in a local shipyard between 1976–84, are in service on the River Danube. Two minelayers of the *'Cosar'* class also double as mine countermeasures support ships.

RUSSIA
Russia currently operates the largest MCM force in the world, but much of it is obsolete.
A number of new designs has been tried in recent years, but none has proved successful. The latest to enter service are the two *'Gorya'* class ocean minehunters completed in 1986 and 1991. A third unit has been seen under construction. Construction of the *'Natya'* class ocean minesweepers ceased in 1980. Currently 27 are in service but early boats are beginning to pay off. Construction continues for the export market at the rate of about one a year, and a large number are in service with overseas navies: India 12; Libya 8; Syria 1; Yemen 1; Ethiopia 1.
A new minehunter based on the *'Natya'* is expected to enter service before the end of the decade. Only two of the *'Yurka'* class ocean minesweepers built between 1963–72 remain operational, and these will have been withdrawn from service by next year. Overseas sales include four to Egypt and two to Vietnam. Of the ubiquitous T 43 class ocean minesweepers built between 1948–57 and of which large numbers were transferred to friendly navies during the Cold War era, only five remain operational, and these will probably have been withdrawn from service within the next two years. In the coastal category of mine warfare vessels construction of the *'Lida'* class minehunter continues at a rate of about two a year. A total of 22 vessels are currently operational. As they enter service older vessels of the *'Vanya'* class minesweepers/hunters (15 currently operational are being paid off for scrapping. Also being removed from service are the *'Yevgenya'* class minehunters (27 units currently operational). Some 62 *'Sonya'* class minesweepers remain operational, but these too are gradually being scrapped. Large numbers of these vessels were also transferred to friendly countries. Only three *'Olya'* minesweepers remain operational, and, like the other obsolete boats, will probably have been scrapped within the next two years. The *'Baltika'* class minesweeper (one unit operational) was a trawler converted to mine warfare in 1983, possibly to test the feasibility of the COOP concept. No other units of this type have been noted. In addition to the aforementioned, Russia also deploys seven drone type craft but this concept does not appear to be continuing. Of the *'Ilyusha'* type only four remain in service and only three of the *'Tanya'* type have been completed. In addition to these units the Russian Navy operates a number of Mi–8 'Hip' and Mi–14 'Haze B' helicopters as well as various other small towed and unmanned craft in the

MCM role. There is also a MCM training and support ship. GRP construction has not proved a success in Russia and the main material used is still wood, sometimes sheathed in GRP.

SPAIN
Plans have been drawn up for Bazan to build a number of minehunters based on GRP technology transferred under a contract signed with Vosper Thornycroft in July 1989. The first unit of a planned four was ordered in May 1993 and laid down in May 1995. it is scheduled to be launched in June 1997. Initially eight minehunters will be built to this new design followed by four sweepers. Spain also operates four ex–American ocean minesweepers of the *'Aggressive'* class and eight coastal minesweepers of the *'Adjutant'/'Redwing'*/MSC 268 classes. These units will gradually be phased out of service as the new CME class enters service.

SWEDEN
A modern MCM force exists based on the seven modern *'Landsort'* class minehunters are now operational; a projected eighth vessel was cancelled. One of the class has been used to carry out trials with the Bofors Sea Trinity CIWS. This self–defence system may be installed in all the units of the class in due course. The *'Landsort'* class are designed to remotely control SAM class influence sweep craft, five of which are operational. Four *'Landsort'* class have also been built for the Singapore Navy. Four units of a new class of inshore minesweepers, the *'Styrso'* class, were ordered in 1994. These craft are designed to control two SAM drones of a new type. A development model will undergo trials next year with an order for eight units anticipated in 1999. These SES type craft will be capable of carrying out a full range of influence sweeps including electric influence sweeps. In addition the Swedish Navy also operates a number of trawler type MCM craft on minesweeping duties. These are now gradually being phased out of service. Four *'Ejdern'* class sonobuoy vessels entered service in 1991 and it is possible that more may be ordered. These craft, equipped with an AQS–928 acoustic processor, are used to lay and monitor sonobuoys in territorial waters to detect intruders. As well as these MCM units, Sweden also operates a transportable COOP system which can be rapidly installed on any type of ship and which is used for route survey tasks.

TURKEY
A major MCM programme is urgently required to replace 11 obsolete ex–American coastal minesweepers. Tenders for the construction of a new class of six MCMVs were put out to France, Germany, Italy, South Korea and the UK in 1991. The bids were delayed until July 1993, and it is not anticipated that any orders will be placed before 1997. Alternatively, if resources cannot be found to fund new construction, a more modern second–hand vessel such as the *'Tripartite'* units of either the Belgian or Dutch navies might possibly be acquired. In addition to the 11 ex–American coastal minesweepers, Turkey also operates six ex–German coastal minesweepers and four ex–American *'Cove'* class inshore minesweepers, all of which are urgently in need of replacement. Eight tender/diver support ships originally built in 1942 are also in service.

UNITED KINGDOM
A major programme of MCM construction continues with the acquisition of the *'Sandown'* class, and orders for a further seven units was placed in July 1994 with the first being laid down in September 1995. Three other units have been built for Saudi Arabia. These vessels are not likely to need a major upgrade until about 2005. A mid–life update for the 13 *'Hunt'* class is scheduled and will include a new variable depth sonar, the Nautis command system and a new mine disposal vehicle – either an upgraded PAP or a AUV craft. It is possible that the new

Australian AMASS influence sweep may also be acquired for the upgrade. In addition to these vessels, the Royal Navy also operates five *'River'* class minesweepers, which also double as patrol craft. The ships were built to carry out deep sweeping operations. With the ending of the Cold War it was considered that these craft, some of which have suffered from corrosion, were surplus to requirements, and as part of a cost cutting exercise four were sold to Bangladesh in 1994 and three to Brazil in 1995. Although the remaining craft are used primarily for patrol duties, and may receive a 30mm in lieu of the 40mm in due course, they may be set aside for disposal in the fairly near future.

YUGOSLAVIA (SERBIA AND MONTENEGRO)
The state of Yugoslavia still has a number of mine warfare vessels under its control. These include two *'Vukov Klanac'* class minesweepers/hunters built in France to the British *'Ton'* design, two British–designed *'Ham'* class inshore minesweepers, seven Yugoslav–designed and built *'Nestin'* class river minesweepers, with possibly another two under construction, and four M 301 class river minesweepers which will have been removed from service by the end of 1996. All but the *'Nestin'* class are now very much overage and in need of replacement. Construction of the *'Nestin'* class may continue, as long as it is able to acquire the necessary materials.

2.2.2 Middle East, North Africa, Gulf & Western Asia

ALGERIA
One ex–Soviet minesweeper is probably non–operational. Plans exist to acquire two new mine warfare vessels before the end of the decade, but this programme may be delayed.

BANGLADESH
In 1994 four ex–UK Royal Navy *'River'* class minesweepers were acquired. In addition Bangladesh is acquiring newly–built T 43 minesweepers from China. Ordered in 1993, one is operational and another three are under construction.

EGYPT
A modern MCM capability is being developed with three GRP–hulled coastal minehunters ordered from Swiftships in the US in 1990, the first unit being delivered in June 1994. Two more were delivered in 1995 and third in 1996. Apart from these new additions the Navy operates a mixed force of six ex–Soviet T 43 type and four *'Yurka'* class ocean minesweepers which are now obsolescent. With the delivery of the new minehunters plans to upgrade the T 43 and *'Yurka'* class ships have been abandoned. Swiftships has also delivered two route survey vessels in 1993, and another two are projected for construction in an Egyptian yard.

INDIA
With an extensive coastline and many major ports and naval bases, not to mention a large merchant fleet, the current MCM capability is inadequate for the task. The force needs to be expanded and existing units either need urgently replacing or modernising with new systems. The need for at least 10 minehunters has been recognised, with construction of six of the units to be undertaken at the Goa shipyard, which is said to be building a GRP facility to cope with the anticipated order, but no decision has yet been made concerning this programme. An alternative would be to build the 10 vessels to a Russian design, possibly the *'Natya'*, 12 of which are already in service, having been built in Russia and transferred between 1978 and 1989. The early units are now very much in need of a major upgrade. India also operates six *'Yevgenya'* class inshore minesweepers delivered between 1983–84.

TABLE 2.3A Mine Warfare Vessel Order of Battle 1946–96
MIDDLE EAST, GULF, NORTH AFRICA & WESTERN ASIA

Region/Country	1946	1956	1966	1976	1986	1996
Algeria	0	0	1	2	2	0
Bangladesh	0	0	0	0	0	5
Egypt	3	14	16	12	6	15
India	11	12	10	8	16	18
Iran	0	1	6	5	5	5
Iraq	0	0	0	5	8	2
Israel	2	0	0	0	0	0
Libya	0	0	2	0	2	8
Morocco	0	0	0	1	0	0
Pakistan	0	5	8	7	3	7
Saudi Arabia	0	0	0	0	4	7
Syria	0	0	2	3	5	10
Tunisia	0	0	0	1	2	0
Yemen	0	0	0	3	0	6
TOTAL	**16**	**32**	**45**	**47**	**53**	**83**

IRAN
No effective minehunting force exists although there is a small force of five obsolete ex-American minesweepers. All these vessels are probably not operational, due to lack of spares.

IRAQ
Two Yugoslav-built *'Nestin'* class remain in the Navy together with two *'Yevgenya'* class inshore minesweepers which are unseaworthy.

KUWAIT
There are long-term plans to acquire three minehunters.

LIBYA
Eight ex-Soviet *'Natya'* ocean minesweepers delivered in the early to mid-80s remain in service and are used primarily on patrol duties.

PAKISTAN
Three French *'Tripartite'* MCMVs are being acquired to replace the two obsolete MSC 268 class coastal minesweepers which will be decommissioned during 1996. The first of these was a unit transferred directly from the French Navy and commissioned in 1992. The second unit built new in France is being delivered in 1996, while the third is being fitted out in Pakistan under a technology transfer agreement. It is due to commence sea trials at the end of 1996. The vessels may be equipped with a towed array minehunting sonar. The Pakistan Navy also operates 5 Chinese-built *'Futi'* drones which are controlled from a shore station.

SAUDI ARABIA
An effective MCM force is being built up with three *'Sandown'* class units which were ordered in 1988 the first sailing for Saudi Arabia in 1995 and the second in 1996. Three more planned units have not yet been ordered. The Navy also has three ex-American MSC 322 class coastal minesweepers/hunters. These will be withdrawn from service when the *'Sandown'* class become fully operational in the Navy. No further mine warfare units are planned and it will be some

120–15 years before the *'Sandowns'* require a major mid–life update.

SYRIA
An MCM capability exists, although much of it must be considered obsolete. All units are of ex-Soviet design. The most modern unit is a *'Natya'* class vessel which has had its MCM gear removed and converted to operate as an intelligence vessel. The status of the one T 43 class ocean minesweeper is uncertain, but it is thought to be unseaworthy. One *'Sonya'* class coastal minesweeper was delivered in 1986 and remains operational. Two *'Vanya'* class coastal minesweepers were delivered in 1973, but probably only one remains operational. Finally five new–build *'Yevgenya'* class inshore minesweepers delivered in the mid–1980s remain in service.

YEMEN
During the 1980s five *'Yevgenya'* class minehunters were delivered to the Navy and in 1991 Russia transferred a *'Natya'* class ocean minesweeper.

TABLE 2.3B Mine Warfare Vessel Order of Battle 1996*
MIDDLE EAST, GULF, NORTH AFRICA & WESTERN ASIA

Type/Country	Bangladesh	Egypt	India	Iran	Iraq	Libya	Pakistan	Saudi Arabia	Syria	Yemen	TOTAL
MSC 292/268				3			2				**3**
MSC 322								4			**4**
T 43	1	6							1		**8**
Cape				2							**2**
Futi (drone)							5				**5**
Natya			12			8			1	1	**22**
Nestin					2						**2**
River	4										**4**
Sandown								3			**3**
Sonya									1		**1**
Swiftships		3									**3**
Swiftships (RS)			2								**2**
Tripartite							2				**2**
Vanya									2		**2**
Yevgenya			6						5	5	**16**
Yurka		4									**4**
TOTAL	5	**15**	**18**	**5**	**2**	**8**	9	**7**	**10**	**6**	**83**

2.2.3 Asia, Pacific & Australasia

AUSTRALIA
The May 1991 Force Structure Review recommended the urgent acquisition of a proven design of coastal minehunter. The Review and subsequent Minehunter Coastal Project, projected a force of six vessels, and in 1994 a consortium of Australian Defence Industries and Intermarine of Italy were awarded the contract for the construction of the *'Gaeta'* type vessels. The first unit was delivered to Australia for outfitting in 1995 and is due to commence sea trials in 1998. The remaining ships in the class will all be built in Australia. The Australian Navy also operates two *'Bay'* class catamarans. The design has not proved entirely successful, and no further units are planned, although it had been hoped that the design might prove to be the forerunner of a large class. Much effort has been devoted to developing the COOP (Craft Of Opportunity) concept. A number of auxiliary minesweepers are in service and equipment is stored ready to equip large numbers of fishing vessels which have already been earmarked for conversion into mine warfare

vessels in an emergency. Three GRP–hulled drones were delivered in 1991–2 for deployment by the three small auxiliary minesweepers already in service as part of the COOP programme.

CHINA

Although there is an MCM force of considerable size, may not be very effective. The bulk of the fleet consists of 34 T 43 class ocean minesweepers, construction of which continues for export. Plans for a new class of minehunters are thought to have been drawn up and construction of a coastal minesweeper, the *Wosao*, was begun in 1986. The current status of this programme is not known. The fact that only one ship has been observed may indicate problems with the design, and may be the reason for the reported plans to acquire up to 38 Italian *'Lerici'* type minehunters. Up to 60 *'Futi'* class remote control drones are also available, although the majority of these are kept in reserve. Finally there is a force of about 50 auxiliary mine warfare craft of numerous types, including trawlers and junks. China would be very vulnerable to a mine threat, and the forces available are totally inadequate to meet any such threat. However, there is little likelihood of the country having to meet a major mining campaign in the foreseeable future.

TABLE 2.4A Mine Warfare Vessel Order of Battle 1946–96 ASIA, PACIFIC & AUSTRALASIA

Country	1946	1956	1966	1976	1986	1996
Australia	32	19	6	3	1	10
Burma/Myanmar	2	0	1	1	0	0
China	0	4	23	25	148?	95
Ceylon/S Lanka	0	1	0	0	0	0
Indonesia	0	18	23	7	2	13
Japan†	0	35	34	26	37	37
Korea (North)	0	0	0	0	0	23
Korea (South)	0	17	14	13	13	14
Malaysia	0	0	12	6	4	4
Philippines	0	2	2	2	0	0
Singapore	0	0	0	2	2	4
Taiwan	0	8	8	14	13	13
Thailand	0	4	5	5	4	12
Vietnam	0	3	3	2	0	15
TOTAL	**34**	**111**	**108**	**106**	**224**	**240**

† The Japanese Maritime Self Defence Force was formed in 1954.

INDONESIA

Two *'Tripartite'* minehunters are operational, but more ambitious plans to build up to 12 units have been shelved because of budget restrictions. The MCM force urgently needs building up, and the acquisition of nine ex–East German *'Kondor II'* class coastal minesweepers partly meets this need, although their priority task is EEZ patrol. Australian mine sweeping equipment is being acquired for these vessels. Two T 43 class ocean minesweepers are also operational, although the primary role of these vessels is also patrol.

JAPAN

Three new ocean–going *'Yaeyama'* class MCMVs have been built, but the programme has now ended indicating that there may have been problems with the design. The first of the 20–unit *'Hatsushima'* class was ordered in 1976, and followed by the almost identical nine units of the *'Uwajima'* class. The early units of the *'Hatsushima'* class are now beginning to be decommissioned. The first two units of a new class of coastal minehunter to replace the *'Takami'* class were ordered in 1995, with a third in 1996. The Navy also operates three MCM support ships. A new class has been authorised with three vessels laid down. These vessels can also double as minelayers.

NORTH KOREA

A modern force of coastal minesweepers, the *'Yukto'* class, has been built.

SOUTH KOREA

A major mine threat exists, and a new MCM programme designed to realise a force of 15 vessels is planned. Six GRP–hulled minehunters of the *'Kang Keong'* class have now been completed. A new 750 tonne combined minehunter/sweeper design is also in hand with orders for up to seven units expected shortly and construction due to commence around the turn of the century. Eight obsolete ex–American coastal minesweepers are also in service, but are expected to pay off towards the end of the decade and as the new 750 tonne class begins to enter service.

MALAYSIA
A major programme of construction of four Italian *'Lerici'* class minehunters has been completed. Plans for a further four units have been delayed in favour of coastal minehunters, but there are no immediate plans to acquire these latter. The existing units will need a major upgrade around the turn of the century, and plans have already been formulated to upgrade the command system.

MYANMAR (BURMA)
It is anticipated that two new Chinese-built T 43 class ocean minesweepers will be bought.

SINGAPORE
Four Swedish *'Landsort'* type MCMVs have been acquired to establish an MCM capability.

TAIWAN
Four MWV 50 offshore oil support vessels were built in Germany in 1990–91 for the Chinese Petroleum Corporation (CPC), and on arrival in Taiwan were converted to coastal minehunters. A total force of 12 vessels is said to be planned. A new class is projected, and tenders are expected to be issued when funds become available. Existing units comprise five obsolete ex-American coastal minesweepers which are in a very poor condition and four *'Aggressive'* class ocean minesweepers delivered in 1995 after undergoing a major overhaul.

THAILAND
Two *'Bang Rachan'* class type M 48 MCMVs were built in Germany and delivered in 1987–88. These vessels will be in need of a major mid-life upgrade in the fairly near future. Thailand has a requirement for a two new coastal minehunters and ITTs (Invitation to Tender) were due in on April 3, 1996, with an order due to be announced in September 1996, and the first unit to be delivered in November 1998 and the second a year later. These units will replace the two *'Bluebird'* class coastal minesweepers which are now obsolete and of limited operational capability. Thailand also operates a MCM support vessel which has a minesweeping capability. Spare minesweeping equipment is carried for the minesweepers. In addition seven MSBs are operational. Two were built in the USA in the mid-1960s and two more were built in Thailand in 1994. It is planned that more will be built in Thailand to replace the older MSBs.

TABLE 2.4B Mine Warfare Vessel Order of Battle 1966*
ASIA, PACIFIC & AUSTRALASIA

Type/Country	Australia	China	Indonesia		Japan	N Korea	S Korea	Malaysia	Singapore	Taiwan	Thailand	Vietnam	TOTAL
K 8												5	**5**
MSB											7		**7**
MSB 07					2								**2**
MSC 268/269							8						**8**
MWV 50											4		**4**
T 43		34	2										**37**
COOP	5												**5**
Drones	3												**3**
Support vessel											1		**1**
Adjutant											5		**5**
Aggressive											4		**4**
Bang Rachan												2	**2**
Bay	2												**2**
Bluebird												2	**2**
Fukue					1								**1**
Futi (drone)			60										**60**
Hatushima					20								**20**
Hayase					1								**1**
Kondor				9									**9**
Landsort										4			**4**
Lerici									4				**4**
Lienyun												2	**2**
Sonya												4	**4**
Souya					1								**1**
Swallow								6					**6**
Tripartite				2									**2**
Uwajima					9								**9**
Wosao			1										**1**
Yaeyama					3								**3**
Yevgenya												2	**2**
Yukto						23							**23**
Yurka												2	**2**
TOTAL	**10**	**95**	**13**	**37**	**23**	**14**		**4**	**4**	**13**	**12**	**15**	**235**

VIETNAM
MCM capability rests almost entirely on units delivered by the former USSR, comprising two *'Yurka'* class ocean minesweepers transferred in 1979, four *'Sonya'* class coastal minesweepers delivered in the late 1980s, two *'Yevgenya'* class inshore minehunters transferred in 1986 and five K 8 class minesweeping boats transferred in 1978. In addition two *'Lienyun'* coastal minesweepers were acquired from China. Only the *'Sonya'* class vessels are capable of dealing with a modern mine threat. Most of the remaining vessels are very over age and need replacing.

2.2.4 North America, South America & Caribbean

ARGENTINA
Six ex–UK *'Ton'* type mine warfare vessels were purchased in 1967. Two were converted into minehunters in the UK in 1968. Although obsolescent, all six vessels are still operational. However, it is doubtful if the current economic situation and military budget will permit replacement or further major modernisation.

BRAZIL
Six *'Schutze'* class coastal minesweepers acquired in the 1970s are in urgent need of major modernisation. This may be undertaken towards the end of the decade, but funds are limited. In view of the size of the navy, number of bases and so on, the strength of the mine warfare force is inadequate. Also a larger type of vessel is needed to cover the coastal littoral. If SSNs are acquired modern minehunters will be essential. Plans may exist for a new class of minesweeper.

CANADA
There current priority is to build up an effective MCMV force based around the MCDV (Maritime Coastal Defence Vessel), and four of the 12 units are now operational. They have a dual role combining MCM and general patrol, operating with modular payloads enabling them to quickly adapt to different tasks. There are also two auxiliary minesweepers, converted from offshore supply vessels in 1991, and six MCM diving tenders. The diving tenders are to be replaced.

CUBA
A number of ex–Soviet units exist, many of which are probably non–operational. The fleet consists of four *'Sonya'* class coastal minesweeper/hunters transferred in the 1980s and 12 *'Yevgenya'* class minehunters transferred between 1977 and 1984.

ECUADOR
Negotiations commenced in 1995 for the transfer of three *'Schutze'* class minesweepers from Germany. This project may collapse due to shortage of funds.

TABLE 2.5A Mine Warfare Vessel Order of Battle 1946–96 NORTH AMERICA, SOUTH AMERICA & CARIBBEAN

Country	1946	1956	1966	1976	1986	1996
Argentina	0	0	0	6	6	6
Brazil	0	0	4	6	6	6
Canada	38	44	6	6	6	12
Cuba	0	0	0	7	14	16
Uruguay	0	0	0	0	0	4
USA	148	185	128	38	22	22
TOTAL	**186**	**229**	**138**	**63**	**54**	**66**

UNITED STATES

The 14 *'Avenger'* class ocean going MCMVs are now all operational and are gradually being modernised with the Nautis M command system and SQQ–32 minehunting sonar. Early problems experienced with the machinery have now been overcome and the class is said to be operating satisfactorily. Two of the class are now permanently station in the Persian Gulf. The 12 *'Osprey'* class coastal minehunters are also coming on stream with eight units operational and four more under construction. In addition the use of MCACs for use in littoral mine countermeasures is being evaluated. In 1991 two SAM class drones were acquired from Sweden for evaluation and an upgraded variant is being developed with Sweden. The prototype of the new drone is to undergo trials in 1997 with production possibly planned for 1999. In addition to the surface forces the US Navy also operates helicopters from amphibious assault ships for MCM operations. Numerous other developments for MCM in the littoral region are underway including a near-term mine reconnaissance system (NMRS) based on an UUV deployed from a submarine torpedo tube. All developments connected with the COOP concept have now been abandoned.

URUGUAY

Four *'Kondor II'* class coastal minesweepers formerly belonging to the East German Navy have been transferred together with their minesweeping equipment, but without any armament.

TABLE 2.5B Mine Warfare Vessel Order of Battle 1996 NORTH AMERICA, SOUTH AMERICA & CARIBBEAN

Type/Country	**Argentina**	**Brazil**	**Canada**	**Cuba**	**Uruguay**	**USA**	**TOTAL**
COOP			2				**2**
Diving tenders			6				**6**
Avenger						14	**14**
Kingston			4				**4**
Kondor					4		**4**
Osprey						8	**8**
Ton	6						**6**
Schutze		6					**6**
Sonya				4			**4**
Yevgenya				12			**12**
TOTAL	**6**	**6**	**12**	**16**	**4**	**22**	**66**

2.2.5 Africa

ANGOLA

Two ex–Russian *'Yevgenya'* class minehunters were acquired in 1987, but are probably in a very poor state of repair.

ETHIOPIA AND ERITREA

In 1992 two minesweepers were acquired from Russia – a *'Natya'* class ocean minesweeper and a *'Sonya'* class coastal minesweeper.

NIGERIA

Two Italian *'Lerici'* units are operational but are in a very poor state of repair and are unserviceable.

SOUTH AFRICA

The four *'Ton'* class units have been substantially rebuilt and modernised with a new MCM system and are now considered to have another 20 years of life. Four coastal minehunters of the *'River'* class have been built to a design based on the German *'Schutze'* class. These will be upgraded with a more modern minehunting system starting in 1996.

TABLE 2.6A Mine Warfare Vessel Order of Battle 1946–1996 AFRICA

Country	**1946**	**1956**	**1966**	**1976**	**1986**	**1996**
Angola	0	0	0	0	0	2
Ethiopia & Eritrea	0	0	0	0	0	2
Ghana	0	0	3	3	0	0
Nigeria	0	0	0	0	0	2
South Africa	0	2	12	10	10	12
TOTAL	**0**	**2**	**15**	**13**	**10**	**18**

TABLE 2.6B Mine Warfare Vessel Order of Battle 1996
AFRICA

Type/Country	Angola	Ethiopia & Eritrea	Nigeria	South Africa	TOTAL
Lerici			2		**4**
Natya		1			**1**
River				4	**4**
Sonya		1			**1**
Ton				8	**8**
Yevgenya	2				**2**
TOTAL	**2**	**2**	**2**	**12**	**18**

Oksoy, the first of class of Norway's minehunter fleet *(H M Steele)*

SECTION 3 - MINE WARFARE VESSEL EXPORT DESIGNS CURRENT INVENTORIES & MARKET PROSPECTS

3.1 Introduction

At present a total of 52 countries possess mine warfare forces, with a small number who have plans to develop such a capability. To date, 14 countries have indigenously developed MCMV designs, and of these nearly all (excepting Japan) compete in the export market. Some, however, are building indigenously to foreign designs and are subject to any licence agreements they may have concluded with the copyright holders of designs.

Licensed construction is currently in progress or will shortly commence in Australia and Pakistan and possibly also in India and Turkey. The one country which is fully capable of licensed construction in either wood or A-magnetic steel, is South Africa. It would only require a limited degree of technology transfer and the setting up of the necessary facilities, of which the country is quite capable, to undertake GRP construction.

TABLE 3.1 Current MCMV Designs

Europe & Scandinavia		Australasia		North America	
Country	**Design**	**Country**	**Design**	**Country**	**Design**
Belgium	?	Australia/Italy	Gaeta/Huon	Canada	Kingston
Denmark	Flyvefisken	Japan	Uwajima	USA	Avenger
Denmark	SAV	Japan	Yaeyama	USA/Italy	Osprey
Germany	Frankenthal	Japan	MSC 07	USA	Swiftships
Germany	Hameln	South Korea	Swallow		
Germany	Troika				
International	Tripartite				
Italy	Lerici/Gaeta				
Norway	Oksoy/Alta				
Russia	Gorya				
Russia	Lida				
Russia	Natya				
Spain	CME				
Sweden	Landsort				
Sweden	SAM				
Sweden	Styrso				
UK	Sandown				

Of the countries currently developing new designs which may be available for export, Spain has developed one based on a technology transfer from the UK and the setting up of GRP construction facilities in collaboration with DCN. Belgium has also developed a new design which may also be available for export. South Korea is understood to be working on a new design which, initially, will only be available for the domestic market. The design will doubtless be available for the expanding export market which is developing in the Asian region. Taiwan, if political considerations and restrictions can be overcome, might also be interested in licensed construction.

The following pages detail the latest designs, some of which are available for export or licensed construction, together with specifications.

3.2 Europe & Scandinavia

BELGIUM

Minesweeper Project
In 1989 Belgium signed an MoU with the Netherlands to build a new generation of GRP-hulled minesweepers. When the Netherlands withdrew from the project in 1993, Belgium continued on its own and authorisation for the acquisition of four units was given in 1994. Work on the first unit is scheduled to commence in 1997 with completion due in 1999.

The design, the contract for which was awarded in November 1990, has been prepared by Beliard Polyship NV and van de Giessen–de Noord Marinebouw. The ships will be equipped with sophisticated minesweeping equipment including wire, acoustic, and electric sweeps, all handled by an A-frame centreline crane at the stern. Part of this package includes the new, sophisticated Sterne M magnetic minesweeping system developed by Thomson Sintra, evaluation of which began in 1995. The Sterne M influence sweep gear will operate in what is known as the Target Simulation Sweeping Mode which is capable of simulating different ship influences including magnetic, acoustic, extremely low frequency electromagnetic (ELFE) and underwater electrical potential (UEP). Sterne M consists of six different bodies, interconnected and towed in array. The equipment is designed to sweep bottom mines which are buried so deep in soft sand that they cannot be detected using normal minehunting methods.

Main propulsion will be by two 1,088 bhp Brons/Werkspoor diesel engines driving twin shafts; and the vessels will be armed with a single 25 mm AA gun on the foredeck, and a machine gun on each side of the aft end of the long forecastle.

Market Prospects
Prospects for this design appeared good when the project was first revealed. However, since then the Netherlands has withdrawn from the joint venture. The other potential partner in the venture was Portugal. Financial restrictions and other more urgent projects prevented Portugal from actively joining in the project. However, Portugal has maintained a watching brief over the progress of the project and before the end of the decade may well decide to acquire the design. Portugal is still urgently in need of a modern MCM capability and once current projects are completed will probably seek to develop an MCM capability. For the time being, however, Belgium will proceed with the project on its own. A major effort will probably be undertaken to secure other markets and potential buyers for the design (which may well turn out cheaper than other MCM designs currently available). It is likely that this will largely be burgeoning Far East navies. Apart from the potential in the Far East region the other major area where such a design would prove of value would be the Gulf where there is an urgent need to develop an indigenous MCM capability. Finally, the South American market has yet to accept the need for modern MCM forces. Here, too, this design may well prove to be suitable for the navies of this region, the design being capable of operating in a dual role both as a patrol vessel and as MCMV. To date Belgium has not achieved any sales of MCMVs in the export market, although it has been a major partner in the *'Tripartite'* design, which has achieved considerable success in the export field. In addition a number of these vessels are now being made available on the second-hand market. Indeed, the Belgian Navy itself may well put some of its vessels up for sale. With its intimate experience of the *'Tripartite'* project, Belgium has an entree into the export market, and may well be able to capitalise on

this for any potential exports of the new design. This will be further enhanced by the fact that the designers and builders of the new vessels have been very closely involved in the *'Tripartite'* project throughout its life.

DENMARK

'Flyvefisken' class

This multi–purpose design built by Danyard A/S, Aalborg, can be fitted out as a missile boat, patrol boat, minelayer, minehunter or minesweeper, as necessary. The weapon fit is accommodated in four standardised containers – one forward and three aft – each measuring 3.00 x 3.50 x 2.50 m. Container slots not utilised are closed by watertight hatches and displacement, draught and trim vary depending on the role. A passive tank stabilisation system has been installed for use at low minehunting/sweeping speeds, when hydraulic drive is used for propulsion in order to minimise the noise signature. At high speeds rudder roll control is used. The hull structure is of GRP sandwich.

Regardless of role, the machinery, sensor fit and command system is standardised. Main propulsion is by a CODAG arrangement of a General Electric LM–500 gas turbine (on the centreline) and two MTU 16V–396–TB94 diesel engines driving wing shafts fitted with Stone Vickers fixed pitch (centre shaft) and controllable pitch (wing shafts) propellers through Allen double/single–reduction gearboxes. A speed of 12 kts is attained on one diesel engine, 20 kts on two diesel engines, and over 30 kts with all engines on line.

Auxiliary slow–speed propulsion is provided by a 500 bhp General Motors 12V–71 diesel engine driving three Rexroth hydraulic pumps powering a 240 shp Rexroth hydraulic motor on the outer gearboxes, and to the transverse bow thrust unit forward. When operating in the diesel mode the centreline shaft is windmilled by the hydraulic motor to reduce drag, with power supplied by the hydraulic pumps driven from the diesel engine gearboxes. All main and auxiliary machinery is remotely operated and monitored from central control stations on the bridge, and from a compartment forward of the engine room.

The sensor fit comprises a Telefunken SystemTechnik TRS–3D/16 multimode G/H–band phased array surveillance radar integrated with a Hazeltine IFF interrogator which is now replacing the Siemens–Plessey AWS–6 G–band in the earlier boats; Terma I–band Scanter navigation radar; two CelsiusTech tracking radars fitted with TV cameras and laser rangefinders; and Racal SABRE ESM and CYGNUS ECM jammer; all linked to a CelsiusTech 9LV–200 Mk3 AIO/weapon control system.

For minehunting a Thomson–Sintra ASM TSM 2054 sidescan sonar is streamed either from the multi–purpose boat itself, or from two remotely controlled 18m, 30–ton GRP unmanned slave craft (SAV), and Bofors Double Eagle ROV for classification and disposal; together with the associated Thomson–Sintra IBIS 43 minehunting command and TSM 2061 tactical systems. Wire, acoustic and electric sweeps – all handled by a centreline crane – will be provided for minesweeping.

Armament comprises an OTO Breda Super Rapid 120 rpm 76 mm gun is mounted forward. Two 12.7 mm machine guns are also mounted. Also fitted are two Sea Gnat 6–barrelled launchers deploying chaff and IR flares.

SAV Class Minehunter Drone

Built by Danyard these Surface Auxiliary Vessels (SAV) are controlled in pairs by the *'Flyvefisken'* class as configured for MCMV operations. The GRP–hull design is based on the design of the Swedish *'Hugin'* class patrol vessels. Special attention has been given to introducing noise quieting technology. The drone craft can deploy a Thomson–Sintra ASM TSM 2054 sidescan sonar deployed from the stern gantry. The craft are controlled from the parent vessel via a Terma link which is used to relay back to the parent vessel the sonar picture and precise position information. The radio link is also used to carry control signals for the manoeuvring of the drone and control of the sidescan sonar. The drones are fitted with a Furuno I–band navigation radar.

Market Prospects

Naval Team Denmark has been very active in promoting the *'Flyvefisken'* design. The modularity of the design which enables it to rapidly change from one role to another, and the advantages which accrue from the use of many common systems throughout the range of roles, should make this design highly popular among navies where budgets are severely restricted. The Danish Navy has acquired a considerable degree of expertise in the operation of this design, acquiring valuable experience in the management of role–change. With the benefit of this experience behind it, Naval Team Denmark is now able to vigorously promote the full potential of this truly modular design. This design too should prove very attractive to many navies in the Far East, particularly those with restricted budgets. Elsewhere, navies who require a rather more powerful vessel for patrol duties than that which many current MCM designs can offer, may see the *'Flyvefisken'* design as one which could meet their requirements for powerful offshore patrol and also be capable of rapid conversion to the MCM task, depending on how the threat scenario develops. South American navies in particular, would find this capability of considerable benefit, restricted as they are by limited financial resources and budget restrictions. To date Denmark has not achieved great success in the export market, mainly being content to develop designs suitable for the domestic market. As such these designs have not, therefore, always been suitable to meet the requirements of many overseas markets. With the *'Flyvefisken'*, however, Denmark has developed a design which, because of its modular role concept, is eminently suitable to meet the needs of many smaller navies around the world.

The SAV, coupled with the *'Flyvefisken'* concept, puts Denmark in a position to further exploit the MCM market. Many navies are now either actively developing or looking at the concept of the remote controlled minesweeping drone. In their latest variants such drones are also being developed to manage minehunting operations on behalf of a parent vessel, themselves controlling an ROV. In this area of mine warfare Denmark is well placed with its *'Flyvefisken'* and SAV remote control vessel to capture a share in this developing market. However, it may not be until towards the end of the decade that such a development will become mandatory for many navies actively seeking to develop modern MCM capabilities.

GERMANY

***'Hameln'* class minesweepers**

The 10 Type 343 minesweepers of the *'Hameln'* class built by Lurssen Werft, Abeking & Rasmussen and Krogerwerft are designed for minesweeping operations in the Baltic with *'Troika'* control capability, and with mechanical, acoustic and magnetic sweeps. The hull and superstructure are completely constructed from non–magnetic and non–corrosive austenitic steel, fully welded. Plates of 4–6 mm thick are used for the hull, welded to web frames at intervals of about 1m with longitudinals at 300 mm. The longitudinal stiffenings are offset bulb plate or flat–bar sections. The transverse web frames are built of girders with single sided web plate. Partitions are made from non–magnetic sandwich construction, elastically mounted for shock resistance. Although aluminium has non–magnetic properties, eddy currents cause difficulty, creating significant magnetic signatures when moving in the earth's magnetic field. By using thinner non–magnetic steel, which has a higher resistivity than aluminium and ordinary steel, this has been overcome. The non–magnetic steel does, however, possess sufficient conductivity that there is no problem in achieving earth bonding and electromagnetic shielding, a problem which has been experienced with completely non–conductive materials such as wood and GRP.

The sensor fit includes a Signaal WM–20/2 I/J–band surveillance and tracking radar; Raytheon SPS–64 I–band navigation radar; Palis data link; and Thomson–CSF DR–2000 ESM. The ships are also fitted with the STN ATLAS Elektronik DSQS–11M hull–mounted mine avoidance sonar. All systems are integrated into the minehunting control and M–20 weapon control system (the latter removed from the *'Zobel'* class torpedo boats). Modernisation plans envisage that all the vessels will, in due course, be fitted with the full DSQS–11M minehunting sonar. Five of the class will be modernised with improved SDG–31 mechanical sweeps and be fitted to control improved *'Troika'* drones.

Main propulsion is by two MTU 16V 396 TB84 diesel engines each rated at 3,070 bhp at 2,000 rpm and driving twin shafts fitted with Escher Wyss controllable pith propellers through Renk single–reduction gearboxes. Electric power at 400 V 3–phase 60 Hz is supplied by three Siemens 230 kW alternators, each powered by a MWM TBD–601–6S auxiliary diesel rated at 639 bhp at 1,800 rpm.

The craft are armed with two Bofors 40 mm AA (2x1) guns, two Stinger SAM quad launchers, and two Silver Dog chaff rocket launchers, and can stow up to 60 mines.

'Frankenthal' class Minehunter

The 10 Type 332 minehunters of the *'Frankenthal'* class, also built by the same builders as the *'Hameln'* class, have the same hull and sensor fit (less the M–20 element) as the *'Hameln'* class; but with the addition of the STN ATLAS Elektronik MWS 80–4 minehunting command system and two Pinguin ROVs for mine disposal in place of the sweep gear. There is no *'Troika'* control or minelaying capability incorporated in these vessels. Main propulsion machinery is identical to the *'Hameln'* class, but with the addition of an electric slow speed drive for minehunting. Two quad Stinger SAM launchers are mounted, but only a single 40 mm is mounted forward. Five of the units are to be modernised to deploy two new *'Troika'* type minehunting drones version. operating towed sidescan sonars and synthetic aperture sonar. In addition they will probably be fitted to operate the new disposable Sea Wolf and Sea Fox ROVs which are being evaluated.

Market Prospects

While Germany has achieved world domination in the submarine export market, the same cannot be said of its efforts in the mine warfare vessel export market. To date only six vessels have been sold abroad, two to Thailand and four to Taiwan. The vessels sold to Thailand in the mid 1980s were a one–off design prepared by Lurssen Werft, while the four vessels sold to Taiwan in 1990–91 were originally built by Abeking & Rasmussen as offshore support vessels for the Chinese Petroleum Corporation, and which on arrival in Taiwan were converted into minehunters. Of the latter it has been mentioned on a number of occasions that the Taiwanese may have plans to build more of these vessels themselves, but to date there has been no confirmation of this and in fact plans have been projected for a new class. Thailand has a requirement for new coastal minehunters. Having already sold two vessels to the country, there is potential for Germany to sell a new design to the Thai Navy, and it is believed that tenders have been put in to meet the new requirement. Germany, however, will face stiff competition for this contract, and the Thais may well decide to place the contract with another company in view of certain difficulties previously experienced with the M 48 design. In other areas Germany has sold on surplus MCMVs to Brazil, and there may be a future market for any vessels surplus to requirements in the German Navy. However, there do not appear to be any moves towards exporting either the *'Hameln'* or *'Frankenthal'* MCMV designs, and the German Navy has probably placed severe restrictions on the sale of these designs.

INTERNATIONAL

'Tripartite' MCMV

The *'Tripartite'* design has been a collaborative venture between France, Belgium and the Netherlands, each nation building its own hulls and with France providing the MCM gear and electronics, Belgium electrical the installation and the Netherlands the engine room equipment.

The hull, decks and partitions are constructed of a single GRP skin stiffened by trapezoid section formers. Hull and former joints are reinforced with fibreglass pins. Hull reinforcements are principally transverse, but longitudinal ribs and binding strakes provide resistance to buckling. The hull is fitted with an active tank stabilisation system, full NBC protection and air conditioning. All the ships are equipped to ship 5 ton container which is stored for variety of tasks such as HQ support, research, patrol, extended diving, and drone control The Indonesian ships exhibit some differences compared with the standard European *'Tripartite'* vessels. Because the ships are expected to take on additional roles as minesweepers and patrol ships deckhouses and general layout are different. The complement varies according to the assigned role and when minehunting six divers are carried.

Two independent propulsion systems are fitted: a main conventional diesel and propeller shaft (supplied by the Netherlands); and an auxiliary propulsion system for minehunting operations which consists of active rudders (supplied by Belgium). The main engine is a supercharged 1,860 bhp (1.37 MW) Stork Werkspoor A–RUB 215 V12 diesel driving, via a flexible coupling and a Rademakers epicyclic reduction gearbox, a Lips 5–bladed controllable pitch propeller. Under auxiliary power the propellers are maintained in a 'feathered' position. The auxiliary propulsion system comprises two Acec active rudders with 6–bladed fixed pitch propellers. In addition a Schottel thrust unit comprising two electric motors each driving a propeller transversely mounted in a tunnel is fitted near the bows. The active rudders and bow thrust unit are driven by three Turbomeca Astazou gas turbine powered alternators mounted high up in the ship. When minehunting one gas turbine unit powers the auxiliary propulsion system, a second provides electrical power for ship's services and the third remains on standby. A fourth diesel–driven alternator provides electric power for ship services when in port. The Indonesian ships are fitted with a different main propulsion system consisting of two 2 MTU 12V 396 TC82 diesels.

The machinery can be controlled from the bridge in manual mode, or from the operations room in either manual or automatic mode. In emergency the machinery can be controlled from a sound–proofed machinery control room located above the engine room.

The minehunting system comprises the Thomson Sintra DUBM 21B hull–mounted sonar integrated with a Thomson CSF Evec data display system. Navigational systems include a Racal Decca 1229 I–band radar, Loran and Syledis and Decca Hifix. Mine disposal is carried out using the two ECA PAP 104 ROVs carried on board. Light mechanical and acoustic sweep gear (the AP–4) can be deployed from the starboard side of the vessels. The Indonesian ships are fitted with a different mine warfare system including the Signaal Sewaco–RI action data automation system; Thomson Sintra Ibis V minehunting system; Thomson Sintra TSM 2022 minehunting sonar; Thomson–CSF Naviplot TSM 2060 tactical display; mechanical sweep gear; and Finnish magnetic and Swedish acoustic sweeps.

The vessels are fitted with comprehensive communications facilities comprising an SNTI internal communication system; two sets of Signaal HF radio receivers; and a UHF receiver supplied by Belgium. Self defence armament comprises a Giat 20F2 20 mm gun (two or three Rheinmetall guns on the Indonesian ships) and a single 12.7 mm machine gun. In addition the Dutch and Indonesian ships can carry a short–range missile system for patrol duties.

Market Prospects

The *'Tripartite'* design has achieved a measure of success on the export market with two units being sold to Indonesia and three to Pakistan. However, it is within Europe that the design has achieved its greatest success in a joint venture between Belgium, France and the Netherlands leading to the construction of 32 vessels. However, success cannot be measured in terms of numbers alone. The design has been in service for almost 13 years, but this did not deter the Pakistan Navy from placing a contract for the acquisition of three *'Tripartite'* hulls in 1992. The first was a recent unit handed over to Pakistan by the French Navy, with a replacement unit being built for the French Navy, and with the second two hulls being built directly for Pakistan. That Pakistan has been willing to acquire a design that is now at least 15 years old, but fitted out with more modern systems, is some measure of the acceptance of the success and capability of the design. The choice by Pakistan to acquire new–build *'Tripartite'* hulls will not go unnoticed by other navies in the region, and will give a boost to any future moves by any of the joint venture partners to export the design. Certainly the Far East will be a region that will be targeted by anyone wishing to export the *'Tripartite'*, and both France and the Netherlands with their previous contacts in the region will be well placed to secure any potential orders. Thailand and Indonesia are both countries that would be major markets for the *'Tripartite'*. However, France is not the only nation that has achieved export success in the region, Sweden now being a major contender for any contracts in the Far East and in the Indian sub–continent region.

ITALY

***'Gaeta'* class**

This is a slightly lengthened version of Intermarine's *'Lerici'* class minehunter and is equipped with SMA MM/SPN–703 radar; Fiar SQQ–14(IT) (an Italian variant of the US SQQ–14 with digital processor) VDS minehunting sonar forward of the bridge; GEC–Marconi Speedscan sidescan route–mapping sonar; one Alenia MIN Mk 2 ROV and one Gaymarine Pluto Plus ROV for mine disposal; and a single Mk 4 mechanical wire sweep. Command and control is exercised through a Datamat MM/SSN–714 command system. Two Oerlikon 20 mm

AA are mounted for self-defence, and when on deployment in threat situations an additional two 20 mm are carried. The improved minehunting system installed in the *'Gaeta'* class was retrofitted to the *'Lerici'* class in 1991.

Main propulsion is by a Fincantieri-GMT BL 230-8M diesel engine driving a single shaft fitted with a KaMeWa controllable pitch propeller through a Tosi single-reduction gearbox. Electric power at 440 V 3-phase 60 Hz is supplied by three Ansaldo 420 kW alternators, each driven by a 550 bhp Isotta-Fraschini ID 36 SS-6V auxiliary diesel engine. For slow speed propulsion there are three (only two in the *'Lerici'* class) 170 shp electro-hydraulic motors each driving a Riva-Calzoni retractable/rotatable thruster positioned one forward and two aft. All machinery is resiliently mounted, and constructed of A-magnetic material.

To prevent buckling, Intermarine developed a monocoque GRP hull of relatively thick single skin whose thickness decreases from keel to deck sides. The material used is Intermarine's patented unsaturated isophthalic resin system which achieves twice the design-lethal limit specified by the Italian Navy and five times the shock resistance of wooden minesweepers. Machinery, equipment and crew are physically isolated from the hull bottom by means of cradles suspended from bulkheads which minimises shock effect on the deck and bulkheads and considerably reduces self-generated noise and vibration, much of which is dissipated before it reaches the hull.

Market Prospects

The two Italian designs, *'Lerici'* and *'Gaeta'*, have achieved major success on the export market. Four *'Lerici'* type MCMVs have been sold to Malaysia, two to Nigeria, and 12 modified *'Gaeta'* design have been built or are building for the US Navy while another six are now under construction for the Royal Australian Navy (see below for details of Australian and US designs). These are in addition to the 12 vessels in service with the Italian Navy. Italy is now the world's leading exporter of MCMVs and will fight hard to retain its position in this field. Future prospects for the design must be good, and Italy will make a major effort to capture any potential contracts in the Far East. Prospects in the Middle East and Gulf are not so certain, but again Italy will make a determined bid to negotiate any future deals with navies in this region. Italy already has an entree into the region, being well known for its small fast patrol boat designs which are in service with a number of navies there. it will, however, face stiff competition from Germany for any contracts in the Gulf for a number of navies in this region have already acquired fast attack craft from the same company in Germany which could supply mine warfare vessels.

NORWAY

'Oksoy/Alta' classes

One of the main features of this SES design, developed by the Norwegian Navy in co-operation with the Defence Research Institute and Norsk Veritas and built by Kvaerner Mandal, is the high transit speed requiring less power than comparable conventional designs, which will facilitate deployment over the lengthy Norwegian coastline. By using a catamaran design built in advanced composite sandwich structure of Fibre Reinforced Plastics (FRP), it is claimed that lower magnetic and acoustic signatures will be exhibited due to the small wet area of the hull, leading to safer operation and less susceptibility to shock, and reduced disturbance of the water leading to improved sonar response, particularly in shallow water. Other advantages claimed are improved manoeuvrability, reduced draught, improved view from the bridge – an important element in minesweeping operations, increased deck area (approximately 70% more compared to a monohull) and greater comfort and improved sea keeping qualities due to a much lower roll acceleration compared to other catamaran type designs and a much reduced roll angle compared to standard displacement hulls (approximately 2–3° in sea state 3).

The minehunter version is equipped with an integrated minehunting sonar comprising the Thomson Sintra TSM–2023N hull–mounted classification sonar and Simrad detection sonar, the combined system being carried on a retractable cantilever amidships inside the air cushion. Two Gaymarine Pluto ROVs are carried for mine disposal carried in a large hangar, being handled by two hydraulic deck cranes. Minehunting operations will be controlled by the Simrad/Thomson Sintra Micos integrated MCM system. The minesweepers are fitted with the Simrad SA 950 hull–mounted mine avoidance sonar, and the customary wire/acoustic/magnetic sweeps which are handled by an A–frame aft. Extensive navigation equipment is fitted including a Racal Decca radar and Doppler log, GPS, wind sensors and so on. These are linked to the Norcontrol Databridge 2000 ARPA and SeaMap electronic chart system.

An Aeromaritime Systembau integrated computerised distributed communications system is fitted, incorporating a comprehensive message handling system. The company is also responsible for the EMI environmental control on the craft.

The ships are armed with a twin–tube Sadral SAM launcher (mounted forward of the bridge) firing Mistral missiles, and one or two Rheinmetall 20 mm AA guns for self–defence.

Main propulsion is by two raft–mounted MTU 12V–396 TE84, 1,920 bhp (1,400 kW) diesel engines powering twin Kvaerner S62/2–80 Eureka water jets through ZF BW 465 single reduction gearboxes. The waterjets are controlled by the dynamic positioning system. Two MTU 8V–396 TE54, 960 bhp (700 kW) auxiliary diesel engines power the lift fans for a draught of 0.90m when on–cushion. Ships services are provided by two MTU 12V 183 TA51 380 bhp generator sets. Machinery and ship control is exercised through a Norcontrol Automation Ship Technical Control system (STC) and a Surface Surveillance System (SSS).

Market Prospects

This is Norway's first venture into the mine warfare vessel design and build market, and in addition it has chosen a revolutionary SES design. While the design itself meets all the criteria for a fully capable mine warfare vessel, the SES concept has not been tried in service over a period of time in this role by any other major navy. In view of this, many navies will want to wait and see how the concept fares in service with the Norwegian Navy before making any decision as to the possibility of acquiring an SES design for their own mine warfare forces.

RUSSIA

***'Gorya'* class**

The hull is probably of aluminium construction. Main propulsion is by two diesel engines driving two shafts for a speed of 18 kts. It is armed with eight (2x4) SA–N–5/SA–N–14 (Grail/Gremlin) SAM missile launchers; one 76.2 mm DP gun; one 6–barrelled 30 mm AA gatling type gun; and two 16–barrelled chaff rocket launchers. The associated sensor fit includes: Palm Frond I–band surveillance radar; Bass Tilt H/I–band fire control radar; Nayada I–band navigation radar; Salt Pot IFF interrogator; HF hull–mounted minehunting sonar; passive ESM (Cross Loop and Long Fold); and a Kolonka optronic director. Accurate position fixing systems may be fitted.

Mechanical, acoustic and magnetic sweeps are provided for minesweeping, and HF sonar and ROV for minehunting. The ROV is deployed from behind a large sliding door sited on 02 deck at the after end of the superstructure to port and starboard beneath the gatling gun. The sweep deck aft is very cramped for space.

'*Lida*' Class
This class of coastal minesweeper was designed as a follow-on to the '*Yevgenya*' class to which it is similar in appearance. The ships are powered by three 900 hp (690 kW) diesels driving three shafts for a speed of 12 knots. The vessels are armed with a 30 mm 6-barrel gatling type gun and minesweeping equipment comprises two acoustic and a single wire sweep. In addition the vessels carry a towed underwater TV camera for mine identification. A Karbarga I hull-mounted high frequency minehunting sonar is carried. The radar outfit comprises a Pechora; I-band surveillance radar which may be complemented by Bass Tilt fire control radar in due course

Market Prospects
The former Soviet Union in the past achieved major success in exporting mine warfare vessels, in particular the ubiquitous T 43 design. However, many of the vessels sold or transferred overseas were supplied to friendly navies and for which market there was, in the main, little or no competition from the outside world. The major exception was India, which has built up a major mine warfare force based on Soviet designs. With the collapse of the Soviet Union, and Russia's entry into a market economy, it will find it hard to compete with other nations on the open market with mine warfare vessels designs which are, by and large, far less capable than those available from other nations. Former friendly countries might continue to favour the acquisition of Russian systems, but these countries are severely restricted by economic difficulties, and as Russia now insists on hard cash or deals favourable to Russia, it is unlikely that export of Russian designs will be in the numbers previously recorded.

SPAIN

CME Class
On 4 July 1989 Bazan signed a technology transfer agreement with Vosper Thornycroft for the shipyard to design a new MCM vessel based on the British '*Sandown*' design. This technology transfer agreement was followed by another agreement signed in November 1993 between DCN and Bazan which provides for personnel at Bazan to be trained in GRP technology. A class of 12 vessels is envisaged, eight minehunters and four minesweepers.

Collectively known as the CME (Contra Minas Espanol) the new design will be fitted with a minehunting system manufactured by FABA-Bazan in co-operation with Inisel and based on the British GEC-Marconi Nautis minehunting command system. The minehunting sonar will be a multifunction high frequency VDS SQQ-32 sonar manufactured by ENOSA under licence from Raytheon. For mine neutralisation the ships will carry two Gaymarine Pluto Plus ROVs.

The vessels will be powered by two MTU 6V 396 TB83 diesels (1,523 bhp) built by Bazan and driving two Voith Schneider propellers for a speed of 14 knots. Two side thrusters will also be mounted.

Armament will comprise a single Oerlikon 20 mm GAM–BO1.

Market Prospects
Although the first unit of this new design has yet to enter service, Spain will undoubtedly make strenuous efforts to recoup some of the investment put into developing the design by marketing a variant. One potential customer for the new design will undoubtedly be Portugal, which has a major requirement to develop a modern MCM force. The other major market which Spain will endeavour to capture will be in South America. In conjunction with Portugal, the two countries would be well matched to capture any market in this region for mine warfare vessels.

SWEDEN

***'Landsort'* class**
This is an extremely compact multi–purpose GRP–hulled minelayer/minehunter/minesweeper design built by Karlskronavarvet. For minesweeping duties the class is equipped with wire, acoustic and electric sweeps; and for minehunting with the Bofors Double Eagle ROV for mine disposal. With the addition of a radio command link the ships can control two slave 18m SAM GRP catamarans fitted with acoustic and electric sweeps. In the minelaying role the vessels are fitted with rails laid along each side of the aft deck, but merging at the stern to form a single track over the transom.

The sensor fit includes: Terma navigation radar; Thomson Sintra TSM 2022 minehunting sonar; TVT–300 optronic director fitted with TV camera and laser rangefinder which feed the CelsiusTech 9MJ–400 command system and the 9LV–100 optronic weapon control system. A British Matilda ESM system is also mounted. A Mains integrated navigation and action data automation system developed by Philips and Racal Decca integrates with the 9MJ–400 command system.

The armament comprises a Bofors 40 mm DP gun, two 7.62 mm GP (2x1) machine guns and four 9–barrelled Saab Dynamics Elma ASW rocket launchers. It is possible that the 40 mm will be replaced in time by the Bofors Trinity 40 mm CIWS system which has been trialled on the *Vinga*. In addition the ships carry the Saab Dynamics RBS–70 MANPADS SAM.

Main propulsion is by four SAAB–Scania DSI–14 diesel engines (1,592 bhp) driving two Voith–Schneider cycloidal propellers with the engines coupled in pairs by an especially quiet clutch cone belt transmission system. Electric power at 440 V 3–phase 60 Hz is supplied by two 180 kW alternators, each driven by a 337 bhp SAAB–Scania DSI–11 auxiliary diesel engine; and a 108 kW alternator driven by an 183 bhp SAAB–Scania DN–11 auxiliary diesel engine. All diesel–alternator sets are mounted on the upper deck to reduce the noise signature; while slow–speed propulsion is achieved by using only one main engine per shaft.

***'Styrso'* Class**
In 1994 Karlskronavarvet and Erisoft AB were awarded a contract to build a new class of MCMV capable of controlling two SAM drones. These ships are also designed for use in an inshore surveillance role.

The main MCM systems comprise two Double Eagle ROVs equipped with Tritech SE 500 sonar and mine disposal charges. The ships themselves are fitted with a Reson mine avoidance sonar and an EG & G side scan sonar system for route surveying tasks. The navigation radar is a Racal Bridgemaster I–band system. All systems integrate with an Ericsson tactical data system. For minesweeping the ships carry an AK–90 acoustic, EL–90 magnetic and mechanical sweeps.

The vessels are powered by two SAAB–Scania DSI 14 diesels; 1,104 hp driving two shafts for a speed of 13 knots. A bow thruster unit is also installed.

There is no main armament, but the ships carry two 12.7 mm mgs.

Market Prospects
Seven *'Landsort'* class vessels are operational with the Royal Swedish Navy and four units have been built for the Singapore Navy. Sweden has now captured a major market in South East Asia, and will make determined efforts to further expand its influence in the region. With Singapore a recipient of Swedish naval equipment, other navies will now look to Sweden as a major potential supplier of naval equipment, and in particular mine warfare vessels.

UNITED KINGDOM

'Sandown' class
The *'Sandown'* class, which is also being acquired by the Royal Saudi Navy, is the third generation of fibre reinforced plastic (FRP) minehunters designed to counter the increasingly sophisticated mine threat. To face anticipated future shock levels new concepts have been realised. Shock waves affect FRP hull structures in two ways – first primary shock waves cause the hull, and particularly the bottom, to undergo high accelerations which can result in delamination and breaking of bonded fixtures. Second, bubble pulse pressures cause the hull to whip which can result in the bottom and main deck panels being severely buckled, possibly leading to a breaking of the ship's back. Any structural connections and their designs therefore play a major part in ship survivability in the face of severe shock waves.

New design techniques have enabled the builders to dispense with the need for through bolting of hull frames previously required to meet exacting shock standards demanded in MCMV construction. The FRP skin of the hull bottom is longitudinally stiffened, with stiffeners laminated to the hull using flexible resin fillets. The main deck is also longitudinally stiffened, and only transverse stiffeners are fitted on the sides of the hull. This has eliminated the need for through bolts and crossovers of structural members resulting in a lighter structure, while at the same time achieving the required strength to meet the exacting shock standards. The main watertight and transverse bulkheads are of corrugated FRP, while all other bulkheads are of FRP sandwich construction with a balsa wood core. The stern is specially designed to reduce slamming and improve conditions for the launch and recovery of the ROV, and results in a much more stable platform for working on the deck.

For the second batch of Royal Navy *'Sandowns'*, Vosper Thornycroft is introducing new construction techniques. including the Scrimp (Seeman Composite Resin Infusion Moulding Process) composite resin injection moulding process. In this process the resin is drawn into a sealed mould under a vacuum. The traditional construction process involves laying up the FRP by hand, with resin applied manually. This scrimp process results in a quality of FRP far surpassing existing FRP construction and will result in a lighter and stronger laminate.

From the safety aspect, the new process will virtually eliminate dangerous styrene fumes. Other improvements in construction are being applied to the modular construction techniques resulting in enhancements to three modules in which outfitting will be maximised prior to module installation in the ships.

Another change to the design includes an increase in size of the propulsors in the second batch (from size 16 – 1.6 m diameter – to size 18) which will enable the vessels to maintain speeds currently attained by operational units, but with more equipment on board. Accommodation arrangements are being remodelled to cater for the inclusion of female personnel in the ship's crew.

Other changes to the design include tropicalisation to enable the units to conduct operations in out–of–area regions such as the Gulf. This will primarily affect habitability standards, and include air conditioning. Finally, a new telephone exchange will be installed together with a 2–man compression chamber. Apart from the new construction techniques, all the changes are relatively minor, but will considerably improve the effectiveness of the ships.

The minehunting system comprises the GEC–Marconi 2093 variable depth minehunting sonar (which is integrated with the ship's position control system) and the Nautis command and control system, combined with highly sophisticated navigation equipments. The sonar can be operated either as a VDS, or locked into the hull for operation in shallow water conditions. Navigation equipment comprises a Kelvin Hughes Type 1007 navigation radar with bright raster display for bridge viewing, and a Kelvin Hughes echo sounder. Other navaids include gyrocompass, AGI log which combines EM log with ground correlation, and Evershed Aeolus wind measuring system which are linked into the command system. Radio fixing aids comprise the Racal Hyperfix and QM14 systems and Decca Navigator Mk 21 and Navstar GPS.

All systems are linked to the command system via a Thorn EMI PDM3 dual redundant serial data highway and embedded microprocessors in each of the equipments. The system architecture devolves data processing functions to individual systems rather than through a central processing unit. This enables each system to exercise effective full local control but at the same time allowing the command to more effectively control and integrate the general functions and operations of the various systems through the command system. This allows various system operators to concentrate their efforts on the immediate task in hand.

Communications comprises HF transmitter, LF/MF/HF receivers, V/UHF transceivers and V/UHF FM transceiver.

Main propulsion is by two Paxman Valenta 6–RP–200EM diesel engines each with a continuous propulsion rating of 755 kW (1,012 bhp) at 1,400 rpm driving two Voith–Schneider 16GS cycloidal propellers through fluid couplings and GEC Marine & Industrial Gears single–reduction gearboxes. The Batch 2 vessels for the Royal Navy are fitted with larger diameter Voith Schneider 18GS propellers (1.8 m in diameter).

The low rotational speed of the Voith–Schneider propellers exhibits little or no cavitation and very low noise, providing infinitely variable thrust all round. The diesels are raft mounted using specially tuned resilient mountings to reduce acoustic signature. Electric power at 400 V 3–phase 60 Hz is provided by three Mawdsley 250 kW alternators, each powered by a 335 bhp Perkins V8–250G air–cooled auxiliary diesel engine. For auxiliary slow–speed propulsion each propeller is belt driven by a 135 shp Mawdsley electric motor, and draw their power from any diesel alternator set. Manoeuvrability is further aided by two Schottel electric powered bow thruster units. To reduce the acoustic signature, which was a design priority, the main engines are raft mounted, the electrical plant is resiliently mounted on the upper deck with air-cooled engines, and hydraulic transmission has been replaced by electric drive in the minehunting mode.

Finally the ship is armed with a MSI Defence Systems Ltd 30 mm gun mounted on the foredeck and two ECA PAP 104 Mk5 ROVs for mine classification and disposal. These are carried on a trolley in the hangar and moved out on rails. A single hangar and wire sweep winch can replace the double hangar fitted in the UK ships, together with an extra crane for handling the minesweeping equipment.

The units ordered by the Royal Saudi Arabian Navy are very similar to the UK vessels except that as they are also to be utilised as patrol vessels the main engines are each uprated to 845 bhp for a slightly higher speed, and an optronic director fitted to control the light AA gun on the fore deck.

Market Prospects
Apart from the five units of Batch 1 which are operational, and with seven more Batch 2 vessels on order, three additional units are being supplied to the Saudi Arabian Navy. With this major export success, the *'Sandown'* design is well placed to meet any future requirements in the Gulf region. Once the vessels are fully operational in the Gulf, other navies in the region will have the opportunity to see the vessel's capabilities for themselves, and the future prospects for sales in the region seem good.

3.3 Asia, Pacific and Australasia

AUSTRALIA

***'Huon'* class**
The monocoque GRP *'Huon'* class coastal minehunters now under construction by Australian Defence Industries in partnership with Intermarine of Italy for the Royal Australian Navy, are being built to a modified *'Gaeta'* design at a purpose-built GRP construction yard in Newcastle.

The minehunting system comprises the GEC-Marconi Nautis-II M command and control system. Data will be fed to the Nautis by a comprehensive range of sensors including a GEC-Marconi Type 2093 VDS minehunting sonar, a Kelvin Hughes Type 1007 I-band navigation radar, ESM, Link 11 and AWADI Prism ESM system. For mine disposal the ships will carry two Bofors Double Eagle ROVs carrying Danish Damdic charges. In addition to their primary minehunting role, the ships will also be fitted with a lightweight mechanical sweep and will also be able to deploy the Australian-developed Mini-Dyad influence sweep. Weapons control is exercised through a Radamec 1000N optronic surveillance system.

Propulsion is provided by a single Fincantieri GMT 230 V8 1,600 kW diesel engine driving a controllable pitch propeller. For precise manoeuvring during minehunting the vessels are fitted with three Italian Riva Calzoni retractable and rotatable thrusters powered by three electro-hydraulic motors driven by three Isotta-Fraschini 18 IF 1300 V-form 350 kW generating sets.

Self defence is provided by a single MSI Defence Systems DS30B 30 mm gun and Super Barricade decoy countermeasures.

Market Prospects
These units are being built to an Italian design, and once the Australian Navy order has been fulfilled, it is possible that a licence to construct the design for other navies may be approved. If this is so then Australia will make a very determined effort to capture a share of the market for mine warfare vessels in the Pacific region.

JAPAN

***'Yaeyama'* class**
Three of these 1,000 tonne wooden-hulled ocean minesweepers have been built by Hitachi and NKK for the Japanese Maritime Self Defence Force. The ships are powered by two Mitsubishi 6NMU-TAI diesel engines, each developing 1,200 bhp (1.76 MW), driving twin shafts fitted with controllable pitch propellers through single-reduction gearboxes for a speed of 18 kts. Precise manoeuvring is aided by a 350 hp (257 kW) bow thruster unit. The ships are being retrofitted with an integrated tactical command system. They deploy the Type S-7 deep water minehunting system and the Type S-8 (SLQ 48) deep water mechanical minesweeping system and are equipped with the Furuno OPS9 I-band surveillance radar; Raytheon SQQ-32 variable depth minehunting sonar and Klein Associates AQS-14 sidescan sonar. Armament consists of a single General Electric JM-61 20 mm Sea Vulcan gatling-type gun.

***'Uwajima'* class**
This wooden-hulled class is a slightly lengthened (increase 2.7m) version of the preceding *'Hatsushima'* class MCMV, the additional length being used to provide improved accommodation facilities. The sensor fit includes a Furuno OPS-18B I-band surveillance radar (OPS 39 in later units) and NEC/Hitachi ZQS 2B (ZQS 3 in later units) high frequency hull-mounted minehunting sonar.

Mine destructor equipment is based on the remote-controlled Type S-7 equipped with counter mining charge. The ships are armed with a single General Electric JM-61 20 mm Sea Vulcan gatling type gun mounted forward.

Main propulsion is by two Mitsubishi 6-NMU-TAI diesel engines each developing 1,400 bhp and driving twin shafts fitted with KaMeWa cp propellers through Mitsubishi single-reduction gearboxes; electric power is provided by one 1,450 kW gas turbine dc generator (for the electric sweep), and two 160 kW ac diesel-alternator sets (for ship services).

MSC 07 Type
This new class of coastal minehunters has been designed to replace the obsolete *'Takami'* class, most of which have now been decommissioned. The design differs markedly from the *'Uwajima'* class and will be much more capable than the current mine countermeasures vessels in service in the JMSDF.

Although similar in layout to the *'Uwajima'* class, the hull has been extended aft to provide additional stowage for mine disposal gear. Following Japanese experience in minesweeping operations in the Gulf during the Iraq-Kuwait war, the ships will be fitted with an integrated mine countermeasures command system (the GEC-Marconi Nautis M command system), a version of the Type 2093 VDS minehunting sonar, and the French ECA PAP 104 Mk 5 ROV. It is believed that the Japanese decision is based on a desire to compare the latest generation British and American command and control and sonar systems. Other systems to be fitted include the Australian ADI Dyad minesweeping system, Furuno OPS 39 I-band surface search radar and a single JM-61 20 mm Sea Vulcan gatling type gun.

The vessels will be powered by two diesels developing 1,800 bhp each driving two shafts for a speed of 14 knots.

Market Prospects
It is most unlikely that any of the Japanese mine warfare designs will be offered for export. All naval construction in Japan is governed by very strict rules and the export of hulls is rigorously opposed.

SOUTH KOREA

'*Swallow*' class
The 'Kang Keong' ('*Swallow*') GRP-hulled class designed and built by the Kangnam Corporation features a Racal Mains 500 integrated AIO; GEC-Marconi Type 193M minehunting sonar; Type 2048 Speedscan sonar; Raytheon I-band navigation radar; Racal Decca plotting system; and two Italian Gaymarine Pluto ROVs. The ships are also fitted with a single sweep gear. Self-defence is provided by a Oerlikon 20 mm light AA gun and two 7.62 mm machine guns. Propulsion comprises two MTU 1,020 bhp diesels driving two Voith-Schneider vertical cycloidal propellers, together with a 102 hp bow thruster for precise manoeuvring.

Market Prospects
Once domestic construction has been fulfilled, and probably not before the end of the decade, it is very likely that South Korea will make a major bid to enter the export market in this region. To what degree any export sales might be achieved will very much depend on the current state of the market in the region. South Korea has a flourishing and fast-developing naval industry and it will be well able to meet naval requirements in the region. Whether other countries will be willing to place contracts with South Korea remains to be seen.

3.4 North America, South America and the Caribbean

CANADA

'*Kingston*' class
The 12 steel-hulled vessels of the multi-role MCDV programme and under construction by St John Shipbuilding, have been designed to combine general patrol duties with MCM tasks. The ships are being built primarily to commercial standards using COTS equipment wherever practicable, with military standards being applied to stability, flood control zones, doors, turning/stopping distances and ammunition spaces.

The design allows the vessels to accommodate one of three modular MCM payloads including a Thomson Sintra deep mechanical minesweeping system; route survey system; and mine disposal deploying ROVs. The heart of the MCM system is the integrated command and control geographic database developed by MacDonald Dettwiler which incorporates their geocoded sonar imagery system. The primary task will be route survey, and, using the geocoded database, it is considered that considerable savings in time will be achieved in minehunting tasks, as only new seabed targets will need to be examined at any one time. For route survey the vessels will deploy a remotely operated survey inspection craft and will tow a sidescan sonar to maintain the up-to-date database of seabed objects. The surface search radar will be a Kelvin Hughes 6000 E/F-band, while a Kelvin Hughes I-band radar will be fitted for navigation. Communications facilities include two each of VHF, UHF and HF, as well as secure voice. The ships will also be fitted with an automated message processing system.

Propulsion is provided by four Wartsila-SACM UD 23V12 2,450 bhp diesels directly driving four Jeumont 715 kW alternators to power two Jeumont 2,000 hp (1,150 kW) electric motors each turning a Lips podded Z-drive azimuth thruster fitted with 5-bladed fixed pitch propeller. Electrical services are provided by a Wartsila-SACM UD19 diesel powering a 300 kW alternator and generator. For emergency use a Wartsila-SACM UD60 105 kW diesel prime mover is fitted.

Self defence is provided by a 40 mm Bofors AA gun and two 12.7 mm machine guns.

Market Prospects
Currently all construction is geared towards meeting the needs of the Canadian Navy. Once this contract has been fulfilled, a determined effort will be made to export this design both for mine warfare purposes and to meet any potential requirement for patrol vessels. Whether Canada will be able to secure any export sales will depend very largely on its ability to match the market forces of other countries already well established in this field, and with already established sales records in foreign countries.

UNITED STATES

'Avenger' class

The 14 units of the *'Avenger'* class built by Peterson Builders Inc and Marinette Marine Corporation are built of White oak, Douglas fir and Alaskan cedar, coated with a thin layer of fibreglass on the outside to take advantage of the low magnetic signature of the wood.

The ships are equipped with an Oropesa SLQ-38 Type 0 Size 1 mechanical wire sweep; SLQ-37(V)3 acoustic/magnetic influence sweeps; EDO ALQ 166 magnetic minesweeping vehicle; two Alliant Techsystems SLQ-48 ROVs for mine classification and disposal; General Electric SQQ-30 sonar (being replaced in all ships by the Raytheon/Thomson Sintra SQQ-32) for minehunting; Mk 116 minehunting control system; Cardion SPS-55 I/J-band surface search radar, and an IFF interrogator; Raytheon SPS-66(V)9; I-band navigation radar; and two 12.7 mm GP (2x1) machine guns mounted for self-defence. The last two ships in the series are being fitted with the British GEC-Marconi Nautis minehunting command system, which is also being retrofitted to the rest of the class. The minehunting system integrates the Unisys SYQ 13 command system and SSN 2 PINS navigation system. Communications are provided by the WSC-3 UHF system and the SRR-1 Satcom system.

The first two units of the class are powered by four Waukesha LN-1616 diesel engines; driving twin shafts fitted with KaMeWa cp propellers through DeLaval twin-input/single-output single-reduction gearboxes. Three more Waukesha LN-161 diesels drive Tech Systems

375 kW alternators supplying electric power at 440 V 3-phase 60 Hz. The third and all subsequent vessels are powered by four 655 bhp Isotta-Fraschini ID36SS-6V-AM diesel engines. For auxiliary slow-speed propulsion two Hansome 200 shp electric motors are coupled to each shaft line. For precise manoeuvring the ships are fitted with a single 350 bhp Omnithruster hydrojet. In addition, there is a Siemens alternator powered by a Solar gas turbine for the electric sweep.

'*Osprey*' class

The *'Osprey'* class built by Intermarine, Savannah and Avondale Industries, is an enlargement of the Italian *'Lerici'*. The *'Osprey'* combines a relatively stiff monocoque GRP hull structure, with frames eliminated and with flexible decks and bulkheads connected with flexible attachment details. These have been designed and tested for underwater explosive shock loads. There are no stiffeners to cause stress concentrations on the hull and the attachment of equipment is minimised and located in low deflection areas.

For minehunting the vessels are fitted with the Raytheon/Thomson Sintra SQQ-32 VDS minehunting sonar deployed from a central well forward and Alliant Techsystems SLQ-48 ROVs for mine classification and disposal; the Raytheon SPS-64(V)9 navigation radar and are armed with two 12.7 mm GP (2x1) machine guns. Command is exercised through a GEC-Marconi Nautis M (US designation SSN-2) command and control system linked with a Unisys SYQ 13 and SYQ 109 integrated navigation and machinery control system.. The SLQ-53 deep sweep is being developed with mechanical and modular influence sweep systems and will be included in the mine warfare equipment at a later date.

Main propulsion is by two Isotta-Fraschini ID36SS-V8 AM diesel engines each of 1,600 bhp, driving two Voith-Schneider cycloidal propellers through single-reduction gearboxes (which eliminate need for forward thrust during station keeping); with electric power at 440 V 3-phase 60 Hz provided by three 300 kW alternators each driven by an Isotta-Fraschini ID36SS auxiliary diesel engine. These latter units also power hydraulic pumps for the two 180 shp hydraulic motors for auxiliary slow-speed propulsion, and a 180 shp transverse bow thrust unit forward. The main engines are mounted on GRP cradles and mounted in acoustic housings.

CMH Minehunter

With FMS funding Swiftships have supplied the Egyptian Navy with a class of three GRP-hulled coastal minehunters. The ships are outfitted with a Thoray/Thomson Sintra TSM 2022 hull-mounted minehunting sonar feeding into an improved SYQ-12 command and data handling system. For minehunting the ships carry a Gaymarine Pluto ROV and can stream a side scan sonar from a deck crane. Sophisticated navigation equipment is installed, including a dynamic positioning system, Sperry I-band navigation radar, GPS and standard line-of-sight navigation systems. Two 12.7 mm mgs are carried for self defence. Propulsion is provided by two MTU 12V 183 TE61 diesels each of 1,068 bhp powering two steerable Schottel propellers. In addition a White Gill 300 bhp thruster is mounted near the bows.

Market Prospects

The only sales success that has been achieved by the USA in recent years in the mine warfare field has been the sale to Egypt of three Swiftships mine warfare vessels and two route survey vessels. It is most unlikely that the 'Avenger' design will be exported, and in fact the major shipyard responsible for building the vessels has now been sold to another enterprise. The smaller 'Osprey' design is based on the Italian minehunter design, but again it is unlikely that this design will be made available for export. Any effort made by the US to sell mine warfare vessels overseas will be based on any indigenous design that may be prepared, and as most of these will be unproved designs, it is unlikely that any foreign navy will seek to acquire such designs. On the other hand, the USA does provide certain foreign friendly navies with FMS funding to have vessels built, and it may be that an American yard will build to a foreign design, mine warfare vessels for a third country. Any such construction will, however, have to be the subject of a licence agreement.

TABLE 3:2 Mine Warfare Vessel Designs – Specifications and Electronics

Class	Navy	Displacement[1]	Dimensions[2]	Command	Sonars	Radar	EW	Hull
MINESWEEPERS								
K 8 (M/S boat)	Vietnam	?/26	16.9 x 3.2 x 0.8	None	None	None	None	
MSB (M/S boat)	Thailand	21/25	15.3 x 4 x 0.9	None	None	None	None	Wood
MSB 07 (M/S boat)	Japan	50/?	22.5 x 5.4 x 1	None	None	OPS-29D	None	Wood
MSC (new minesweepers)	Belgium	?/644	52.4 x 10.4 x 3.1		Active	ARPA	ESM	GRP
MSC (river minesweepers)	Romania	?/97	33.3 x 4.8 x 0.9		None	Nayada		
MSC 268	South Korea	320/370	43 x 8 x 2.6	None	UQS-1 or TSM 2022	Decca 45	None	Wood
MSC 268	Pakistan	330/390	43.9 x 8.5 x 2.6	None	UQS-1D	Decca 45	None	Wood
MSC 268 & 292	Iran	320/384	44.5 x 8.5 x 2.5	None	UQS-1D	Decca 707	None	Wood
MSC 289	South Korea	315/380	44.3 x 8.3 x 2.7	None	UQS-1 or TSM 2022	Decca 45	None	Wood
MSC 294	Greece	320/370	43.3 x 8.5 x 2.5		UQS-1D	Decca		Wood
M 15	Sweden	70/?	27.7 x 5 x 2	None	None	Terma	None	
M 301	Yugoslavia	?/38				Racal Decca		
PO 2 Type 501	Bulgaria	?/56	21.5 x 3.5 x 1					
T 43	Albania	500/580	58 x 8.4 x 2.1	None	Stag Ear	Ball End, Neptun	None	Steel
T 43 (China built)	Bangladesh	520/590	60 x 8.8 x 2.3	None	Tamir II	Fin Curve	None	Steel
T 43 Type 010	China	520/590	60 x 8.8 x 2.3	None	Tamir II	Fin Curve/Type 756	None	Steel
T 43	Egypt	500/580	58 x 8.4 x 2.1	None	Stag Ear	Don 2	None	Steel
T 43	Indonesia	500/580	58 x 8.4 x 2.1	None	Stag Ear	Decca 110	None	Steel
T 43 Type 254	Russia	520/590	60 x 8.4 x 2.3	None	Stag Ear	Ball End & Don 2 or Spin Trough		Steel
T 43	Syria	500/580	60 x 8.4 x 2.1	None	Stag Ear (MG 11)	Ball End, Don 2	None	Steel
T 301	Albania	146/170	38 x 5.7 x 1.6	None	None	None	None	Wood
T 301	Romania	145/170	38 x 5.7 x 1.6	None	None	None	None	Wood
Adjutant & MSC 268	Spain	355/384	43.9 x 8.5 x 2.5	None	UQS-1D	TM 626 or RM 914	None	Wood
Adjutant & MSC 268	Taiwan	320/375	43.9 x 8.5 x 2.4		UQS 1D	Decca 707		Wood
Adjutant, MSC 268 & MSC 294	Turkey	320/370	43 x 8 x 2.6	None	UQS-1D	TM 1226	None	Wood
Aggressive	Belgium	720/780	52.6 x 10.7 x 4.3	None	SQQ-14	Type 1229	None	Wood
Aggressive	Spain	720/817-853	52.6 x 10.7 x 4.3	None	SQQ-14	TM 1226 & Decca 626	None	Wood
Aggressive	Taiwan	720/780	52.6 x 10.7 x 4.3		SQQ-14	SPS-53L		Wood
Alta	Norway	?/375	55.2 x 13.6 x 2.5	Micos	SA 950	Racal Decca	None	FRP
Arko	Sweden	285/300	44.4 x 7.5 x 3	None	None	Skanter 009	None	Wood
Baltika Type 1380	Russia	210/235	25.4 x 6.8 x 3.3	None	None	Spin Trough	None	Wood
Bluebird	Denmark	350/376	45 x 8.5 x 2.6	None	UQS-1D	NWS 3	None	Wood

Class	Navy	Displacement[1]	Dimensions[2]	Command	Sonars	Radar	EW	Hull
Bluebird	Thailand	317/384	44.3 x 8.2 x 2.6	None	UQS–1D	Decca TM 707	None	Wood
Cape	Iran	200/239	33.9 x 7 x 2.4	None	None	Decca 303N	None	
Cove	Turkey	180/235	34 x 7.1 x 2.4	None	None		None	
Dokkum	Netherlands	373/453	46.6 x 8.8 x 2.3	None	None	TM 1229C	None	Wood
Frauenlob Type 394	Germany	?/246	38 x 8.2 x 2	None	None	Kelvin Hughes 14/9	None	
Gassten	Sweden	120/135	24 x 6.5 x 3.5	None	None	Skanter 009	None	GRP
Gilloga	Sweden	110/130	22 x 6.5 x 3.5	None	None	Skanter 009	None	Wood
Ham	Yugoslavia	120/159	32.5 x 6.5 x 1.7	None	None	Decca 45	None	Wood
Hameln Type 343	Germany	590/635	54.4 x 9.2 x 2.5	MWS–80[¥]	DSQS–11M[¥]	WM20/2, SPS–64	DR 2000	Steel
Hisingen	Sweden	130/150	24 x 6.5 x 3.5	None	None	Skanter 009	None	Wood
Kiiski	Finland	18/20	15.2 x 4.1 x 1.2	None	None	Type 1229	None	GRP
Kondor II Type 89	Indonesia	?/310	56.7 x 7.5 x 2.4	None	AQS 17	TSR 333	None	
Kondor II Type 89.2	Latvia	?/310	56.7 x 7.5 x 2.4	None	None	I–band	None	
Kondor II	Uruguay	?/310	56.7 x 7.5 x 2.4	None	None	TSR 333 or Raytheon 1900	None	
Krogulec Type 206F	Poland	?/503	58.2 x 7.7 x 2.1	None	MG 11 & SHL 200	TRN 823	None	
Kuha	Finland	?/90	26.6 x 6.9 x 2	None	None	Type 1229	None	GRP
Leniwka Type 410S	Poland	195/269	25.8 x 7.2 x 2.7	None	None	SRN 311	None	
Lienyun	Vietnam	?/400	40 x 8 x 3.5					
Musca	Romania	660/790	60.8 x 9.5 x 2.8		Hull	Krivach, Drum Tilt		
Natya I Type 266M	Ethiopia & Eritrea	650/804	61 x 10.2 x 3	None	Hull–mounted	Don 2	None	Aluminium/ steel
Natya I Type 266M	India	650/804	61 x 10.2 x 3	None	MG 69/79	Don 2, Drum Tilt	None	Aluminium/ steel
Natya I Type 266ME	Libya	650/804	61 x 10.2 x 3	None	Hull–mounted	Don 2, Drum Tilt	None	Aluminium/ steel
Natya I Type 266M	Russia	650/804	61 x 10.2 x 3	None	MG 79/89 or Tamir	Drum Tilt & Don 2 or Low Trough	None	Aluminium/ steel
Natya II Type 266DM	Russia	650/804	61 x 10.2 x 3		MG 79/89	Drum Tilt & Don 2 or Low Trough		Aluminium/ steel
Natya I Type 266M	Syria	650/804	61 x 10.2 x 3	None	?MG 79/89	Don 2, Drum Tilt	None	Aluminium/ steel
Natya I Type 266ME	Yemen	650/804	61 x 10.2 x 3	None	MG 69/79	Don 2	None	Aluminium/ steel
Nestin (river)	Hungary	66/72.3	27 x 6.5 x 1.2	None	None	Type 101	None	
Nestin (river)	Iraq	65/72	27 x 6.5 x 1.2	None	None	Decca 101	None	
Nestin (river)	Yugoslavia	65/72	27 x 6.3 x 1.6	None	None	1226		
Notec Type 207P	Poland	208/225	38.3 x 7.2 x 1.8	None	MG 89 or MG 79	SRN 302	None	GRP

Class	Navy	Displacement[1]	Dimensions[2]	Command	Sonars	Radar	EW	Hull
Olya Type 1259	Bulgaria	44/64	25.8 x 4.5 x 1	None	None	Pechora	None	Wood
Olya Type 1259	Russia	44/64	25.8 x 4.5 x 1	None	None	Don 2	None	Wood
River	Bangladesh	?/890	47.5 x 10.5 x 2.9	None	None	TM 1226C	None	Steel
River	UK	?/770	47.5 x 10.5 x 2.9	None	None	TM 1226C	None	Steel
Schutze	Brazil	230/280	47.2 x 7.2 x 2.1	None	None	ZW06		Wood
Sonya Type 1265	Bulgaria	350/450	48 x 8.8 x 2	None	MG 69/79	Kivach	None	Wood & GRP
Sonya Type 1265	Cuba	350/450	48 x 8.8 x 2	None	MG 69/79	Don 2	None	Wood & GRP
Sonya Type 1265	Ethiopia & Eritrea	350/400	48 x 8.8 x 2	None	MG 69/79	Don 2	None	Wood & GRP
Sonya Type 1265	Syria	350/450	48 x 8.8 x 2	None	MG 69/79	Don 2	None	Wood & GRP
Sonya Type 1265	Vietnam	350/400	48 x 8.8 x 2	None	MG 69/79	Nayada	None	Wood & GRP
Styrso	Sweden	?/175	36 x 7.9 x 2.2	Ericsson	Reson & EG & G	Bridgemaster		
Ton	South Africa	360/440	46.6 x 8.8 x 2.5	None	Type 193**	Type 978 or 1006	None	Wood
Vanya Type 257D	Bulgaria	220/245	40 x 7.3 x 1.8	None	MG 69/79	Don 2	None	Wood & GRP
Vanya	Syria	220/245	40 x 7.3 x 1.8	None	MG 69/79	Don 2	None	Wood & GRP
Vegesack	Turkey	362/378	47.3 x 8.6 x 2.9	None	Simrad	Decca 707	None	
Wosao	China	320/?	44.8 x 6.8 x 2.3		Hull-mounted	China Type 753		Steel
Yevgenya Type 1258	Bulgaria	77/90	24.5 x 5.5 x 1.4	None	MG 7	Spin Trough	None	GRP
Yevgenya	India	77/90	24.6 x 5.5 x 1.5	None	MG 7	Spin Trough	None	GRP
Yevgenya	Syria	77/90	24.6 x 5.5 x 1.5	None	MG 7	Spin Trough	None	GRP
Yukto I	North Korea	?/60	24 x 4 x 1.7	None	None	Skin Head	None	Wood
Yukto II	North Korea	?/52				Skin Head		Wood
Yurka Type 266	Egypt	400/540	52.4 x 9.4 x 2.6	None	Stag Ear	Don 2	None	Aluminium/ steel
Yurka Type 266	Russia	400/540	52.4 x 9.4 x 2.6	None	Stag Ear	Drum Tilt & Don 2 or Spin Trough	Watch Dog	Aluminium/ steel
Yurka Type 266	Vietnam	400/540	52.4 x 9.4 x 2.6	None	Stag Ear	Don 2 & Drum Tilt	None	Aluminium/ steel
MINEHUNTERS								
CME	Spain	?/530	54 x 10.7 x 2.2	Nautis	SQQ-32	I-band		GRP
MHC (Swiftships type)	Egypt	178/203	33.8 x 8.2 x 2.3	SYQ-12 mod	TSM 2022	Sperry	None	GRP
Bay	Australia	100/178	30.9 x 9 x 2	MWS80-5	DSQS-11M	Type 1006	None	GRP sandwich
Circe	France	460/510	50.9 x 8.9 x 3.4	None	DUBM 20B	Type 1229	None	Wood

Class	Navy	Displacement[1]	Dimensions[2]	Command	Sonars	Radar	EW	Hull
Flyvefisken	Denmark	?/450	54 x 9 x 3					GRP
Frankenthal Type 332	Germany	590/650	54.4 x 9.2 x 2.5	MWS-80	DSQS-11M	SPS-64	DR 2000	Steel
Gorya Type 1260	Russia	950/1,130	66 x 11 x 3.3		Hull	Palm Frond, Nayada Bass Tilt	Cross Loop, Long Fold	Wood
Huon (Gaeta)	Australia	?/720	52.5 x 9.9 x 3.0	Nautis 2M	Type 2093	1007	Prism	GRP
Lida Type 1259.2	Russia	?/135	31.5 x 6.5 x 1.6		Karbarga I	Pechora		
Oksoy	Norway	?/375	55.2 x 13.6 x 2.5[3]	Micos	TSM 2023N	Racal Decca	None	FRP
Osprey	USA	750/918	57.3 x 11 x 2.9	SYQ 13 & SYQ 109	SQQ-32	SPS-64(V)9		GRP
Sandown	Saudi Arabia	450/480	52.7 x 10.5 x 2.1	Nautis M	Type 2093	Type 1007	Shiploc	GRP
Sandown	UK	450/484	52.5 x 10.5 x 2.3	Nautis M	Type 2093	Type 1007		GRP
Swallow	South Korea	470/520	50 x 8.3 x 2.6	Mains 500	193M Mod 1 or 3	Raytheon; I-band		GRP
Tripartite	Belgium	562/595	51.5 x 8.9 x 2.5		DUBM 21B	Type 1229C		GRP
Tripartite	France	562/595	51.5 x 8.9 x 2.5	None	DUBM 21B or 21D	Type 1229C	None	GRP
Tripartite	Indonesia	502/568	51.5 x 8.9 x 2.5	Sewaco-RI	TSM 2022	Type 1229C		GRP
Tripartite	Netherlands	562/595	51.5 x 8.9 x 2.6	Sewaco IX	DUBM 21A	Type 1229C		GRP
Tripartite	Pakistan	562/595	51.5 x 8.9 x 2.9	TSM 2061[4]	DUBM 21B or 21D[5]	Type 1229C	None	GRP
Yevgenya Type 1258	Angola	77/90	24.5 x 5.5 x 1.4	None	MG-7	Don 2	None	GRP
Yevgenya Type 1258	Cuba	77/90	24.6 x 5.5 x 1.5	None	MG-7	Don 2	None	GRP
Yevgenya Type 1258	Russia	77/90	24.6 x 5.5 x 1.5	None	MG-7	Spin Trough or Mius	None	GRP
Yevgenya Type 1258	Vietnam	77/90	24.6 x 5.5 x 1.5	None	MG 7	Spin Trough	None	GRP
Yevgenya Type 1258	Yemen	77/90	24.6 x 5.5 x 1.5	None	MG 7	Spin Trough	None	GRP
MINEHUNTERS/SWEEPERS								
MSC 07	Japan	?/510	57.7 x 9.4 x 4.2	Nautis-M	Type 2093	OPS-39		Wood
MSC 322	Saudi Arabia	320/407	46.6 x 8.2 x 2.5	None	SQQ-14	SPS-55	None	Wood
MWV 50	Taiwan	?/500	49.7 x 8.7 x 3.1	Ibis V	TSM-2022	I-band		
Adjutant	Greece	330/402	44.2 x 8.5 x 2.4	None	SQQ-14 or UQS-1D	Decca or 3RM 20R	None	Wood
Avenger	USA	1,145/1,312	68.3 x 11.9 x 3.5	Nautis M[6]	SQQ-30[7]	SPS-55 & SPS-66(V)9		Wood & GRP
Bang Rachan	Thailand	390/444	48 x 9.3 x 2.7	MWS 80R	DSQS-11H	STN 8600 ARPA		Compsite
Gaeta	Italy	592/697	52.45 x 9.9 x 2.9	None	SQQ-14(IT)	SSN-714V(2)	SPN-728V(3)	GRP
Hatsushima	Japan	440/510	55 x 9.4 x 2.4		ZQS 2B	OPS-9		Wood
Hunt	UK	615/750	60 x 10 x 3.4	CAAIS DBA 4	Type 193M Mod 1 Type 2059 & Mil Cross	Type 1006 or 1007	Matilda UAR 1 Mentor A (in some)	GRP
Kingston	Canada	?/962	53.3 x 11.3 x 3.4		Towed side-scan	Type 6000		Steel
Landsort	Singapore	270/360	47.5 x 9.6 x 2.3	IBIS V	TSM 2022	Terma	Nonc	GRP

Class	Navy	Displacement[1]	Dimensions[2]	Command	Sonars	Radar	EW	Hull
Landsort	Sweden	270/360	47.5 x 9.6 x 2.3	9MJ400	TSM–2022	Terma	Matilda	GRP
Lerici	Italy	470/572	50 x 9.9 x 2.6	None	SQQ–14(IT)	SSN–714V(2)	SPN–728V(3)	GRP
Lerici	Malaysia	470/572	51 x 9.9 x 2.8	None	TSM 2022	Decca 1226	None	GRP
Lerici	Nigeria	470/540	51 x 9.9 x 2.8	IBIS V	TSM 2022	Type 1226	None	GRP
Lindau Type 331	Germany	388/463	47.1 x 8.3 x 3	None	DSQS–11 or 193M	14/9 or TRS N	None	Wood
Lindau Type 351	Germany	390/465	47.1 x 8.3 x 2.8	None	DSQS–11	TRS N	ESM	Wood
Notec II Type 207M	Poland	208/225	38.3 x 7.2 x 1.6	None	SHL 100/200	SRN 401XTA or RN 231	None	GRP
River	South Africa	?/380	48 x 8.5 x 2.5	None	Simrad, Klein	Racal Decca		Wod
Sonya Type 1265/1265M	Russia	350/450	48 x 8.8 x 2	None	MG 69/79	Don 2 or Krivach or Nayada	None	Wood & GRP
Ton	Argentina	360/440	46.6 x 8.8 x 2.5		Type 193[8]	Decca 45		Composite
Uwajima	Japan	490/586	58 x 9.4 x 2.9		ZQS 3	OPS–39		Wood
Vanya Type 257D/DM/DT	Russia	200/245	40 x 7.3 x 1.8	None	MG 69/79	Don 2 or Don Kay	None	Wood
Vukov Klanac (Ton)	Yugoslavia	365/424	46.4 x 8.6 x 2.5	None	TSM 2022	DRBN 30		Wood
Yaeyama	Japan	1,000/1,275	67 x 11.8 x 3.1		SQQ–32	OPS–39		Wood
DRONES								
MSD	Australia		7.3 x 2.8 x 0.6	None	None	None	None	GRP
SAM	Sweden	?/20	18 x 6.1 x 1.6	None	None	None	None	GRP
SAM II	Sweden	?/56	22.5 x 11.7	None				
SAV	Denmark	32/38	18.2 x 4.75 x 1.2	None	TSM 2054	Furuno	None	GRP
Futi Type 312	China	47/?	20.9 x 3.9 x 2.1	None	None	None	None	
Futi Type 312	Pakistan	47/?	20.9 x 3.9 x 2.1	None	None	None	None	
Ilyusha Type 1253	Russia	?/85	26.4 x 5.9 x 1.4	None	None	Spin Trough		
Tanya Type 1300	Russia	?/73	26.5 x 4 x 1.5)	None	None	Spin Trough		
Troika	Germany	99	26.9 x 4.6 x 1.4	None	None	None	None	Wood
MISCELLANEOUS								
AN–2 (mine warfare/patrol)	Hungary	11.5	13.4 x 3.8 x 0.6	None				Aluminium
MCM (diving tenders)	France	375/505	41.6 x 7.5 x 3.8	None	None	Type 1226	None	
MCS (training & support)	Russia	?/1,880	59 x 13 x 4.7		None	Palm Frond, Don 2		
MSA(T) (auxiliary M/S tugs)	Australia	?/412	29.6 x 8.5 x 3.4	None	None	Furuno 7040D		
MSA(S) (Brolga)	Australia	?/268	28.4 x 8.1 x 3.5	None	Klein side scan			
MSA(S) (auxiliary M/S)	Australia	?/119	21.9 x 6.4 x 3			I–band		

Class	Navy	Displacement[1]	Dimensions[2]	Command	Sonars	Radar	EW	Hull
MSA (auxiliary minesweepers)	Canada	1,196/2,205	58.2 x 13.1 x 5.2	None	Side scan towed VDS	Brighttrack 502		
MSI (Swiftships route survey)	Egypt	?/ 165	27.4 x 7.6 x 2.4	SYQ–12 mod	EG & G side scan	Furuno 2020		
MSR (minesweeping launches)	Greece	?/21	15.2 x 4 x 1.3	None	None	None	None	
MST/ML (minesweeper support)	Japan	5,600/?	141 x 22 x 5.4					
MST (support ship)	Thailand	1,000/?	55.7 x 10 x 3.1			Decca 1226		
YAG (minehunting tenders)	Turkey	70/?	21.8 x 4.2 x 2.6	None				
YDT (diving tenders)	Canada	70	22.9 x 5.6 x 2.6	None		Racal Decca		Wood
YDT (diving tenders)	Canada	110	27.3 x 6.2 x 2.6	None	Side–scan	Racal Decca		Steel
Type 742 (diver support ship)	Germany	775/825	63 x 10.6 x 2.7	None	None			Wood
Antares (route survey)	France	?/340	28.3 x 7.7 x 4	None	2 x DUBM 41B	Type 1229	None	
Cosar (support ship/minelayer)	Romania	?/1,450	79 x 10.6 x 3.6		Tamir II	Strut Curve, Don 2 Muff Cob, Drum Tilt	Watch Dog	Steel
Ejdern (sonobuoy craft)	Sweden	?/36	20 x 4.8 x 1.3	None		Terma		GRP
Fukue (support ship)	Japan	380/530	52 x 8.8 x 2.4		ZQS 2	OPS–9		
Hayase Class (MST/ML)	Japan	2,000/3,050	99 x 14.5 x 4.2		SQS–11A & ZQS–1B	OPS–14 & OPS–16		Steel
Souya Class (MST/ML)	Japan	2,150/3,300	99 x 15 x 4.2		SQS–11A & ZQS 1B	OPS–14 & OPS–16		Steel

1 Standard/Full Load tons
2 Length (overall) x beam x draught in metres
3 0.9 m on cushion
4 In last pair
5 TSM 2054 in last pair
6 In last two ships and to be retrofitted to rest of class
7 SQQ–32 in MCM 10 onwards and being retrofitted to rest of class
8 In minehunters
¥ Only in .....
** Only in Kimberley

TABLE 3:3 Mine Countermeasure Vessel Designs – Mine Warfare Systems

Class	Navy	ROVs	Sweep Gear	Guns[¶]	Missiles	Mines[†]
MINESWEEPERS						
K 8 (M/S boat)	Vietnam					
MSB (M/S boat)	Thailand					
MSB 07 (M/S boat)	Japan					
MSC (new minesweepers)	Belgium	2 ROVs	Sterne	1 – 25		
MSC (river minesweepers)	Romania					6
MSC 268	South Korea			1x2 –20 or 2x1 –20		
MSC 268	Pakistan			1x4 –23 or 1 –20		
MSC 268 & 292	Iran		Wire + Ma, Ac	1x2 –20		
MSC 289	South Korea			1x2 –20 or 2x1 –20		
MSC 294	Greece			1x2 –20		
M 15	Sweden					
M 301	Yugoslavia			2 –20		
PO 2 Type 501	Bulgaria					
T 43	Albania			2x2 –37		16
T 43 (China built)	Bangladesh		MPT–3 wire Ma, Ac	2x2 – 37 & 2x2 – 25		12–16
T 43 Type 010	China		MPT–3 wire Ma, Ac	1 or 2x2 – 37 1 – 85 in some		12–16
T 43	Egypt			2x2 – 37		20
T 43	Indonesia			2x2 – 37		
T 43 Type 254	Russia			2x2 – 37		16
T 43	Syria			1x2 – 37		16
T 301	Albania			2 – 37		18
T 301	Romania			2x1 – 37		18
Adjutant & MSC 268	Spain			1x2 – 20		
Adjutant & MSC 268	Taiwan			1x2 – 20		
Adjutant, MSC 268 & MSC 294	Turkey			1x2 – 20		
Aggressive	Belgium					
Aggressive	Spain	Pluto[$]		1x2 –20		

Class	Navy	ROVs	Sweep Gear	Guns[¶]	Missiles	Mines[†]
Aggressive	Taiwan		SLQ-37 wire, Ma & Ac sweeps			
Alta	Norway	None	Wire + M, A	1 or 2 – 20	1x2 Sadral	
Arko	Sweden			1 – 40		
Baltika Type 1380	Russia					
Bluebird	Denmark			1 Bofors 40		
Bluebird	Thailand		Mk 4 & Mk 6 Type Q2 Ma	1x2 – 20		
Cape	Iran		Wire, Ma, Ac			
Cove	Turkey					
Dokkum	Netherlands			1 or 2 – 20		
Frauenlob Type 394	Germany			1 – 40		?
Gassten	Sweden			1 – 20		
Gilloga	Sweden					
Ham	Yugoslavia			1x2 – 20		
Hameln Type 343	Germany		SDG-31 wire	2x1 – 40	2x4 Stinger	60
Hisingen	Sweden			1 – 20		
Kiiski	Finland					
Kondor II Type 89	Indonesia		Dyad influence	3x2 – 25		2 rails
Kondor II Type 89.2	Latvia	None	Temp. removed	3x2 – 23		
Kondor II	Uruguary		MSG-3 variable depth	1 – 40		2 rails
Krogulec Type 206F	Poland			3x2 -25	2x4 SA-N-5	16
Kuha	Finland	1 x Pluto	M,A, P	1x2 -23		
Leniwka Type 410S	Poland		Wire+M,A			
Lienyun	Vietnam					
Musca	Romania			2x2 – 30		
Natya I Type 266 M	Ethiopa & Eritrea			2x2 -30 & 2x2 – 25		10
Natya I Type 266M	India		GKT-2 wire AT-2 Ac TEM-3 Ma	2x2 – 30 & 2x2 – 25	2x4 SA-N-5[$]	10
Natya I Type 266ME	Libya		GKT-2 wire AT-2 Ac TEM-3 Ma	2x2 – 30 & 2x2 -25		10
Natya I Type 266M	Russia		1 or 2 GKT-2 wire	2x2 -30 or 2x6 -30 &	2x4 SA-N-5/8	10

Class	Navy	ROVs	Sweep Gear	Guns[¶]	Missiles	Mines[†]
			1 AT–2 Ac 1 Tem–3 Ma	2x2 –25		
Natya II Type 266DM	Russia			2x2 – 30 or 2x6 – 30	2x4 SA–N–5/8	10
Natya I Type 266 M	Syria			2x2 – 30 & 2x2 – 25	2x4 SA–N–5	
Natya I Type 266 ME	Yemen		Wire,Ma,Ac	2x2 – 30 & 2x2 – 25		10
Nestin (river)	Hungary	None	M,A Kram wire	1x4 –20, 2x1 –20		24
Nestin (river)	Iraq		Wire+Ma,Ac	1x3 –20 & 2x1 – 20		24
Nestin (river)	Yugoslavia		Wire+M, A	1x3 – 20 & 2x1 – 20		24
Notec Type 207P	Poland		Wire+M, A	1x2 – 23		24
Olya Type 1259	Bulgaria		AT–6, SZMT–1 3 PKT–2			
Olya Type 1259	Russia			1x2 – 25		
River	Bangladesh		Wire	1 – 40		
River	UK		Wire	1 – 40		
Schutze	Brazil		Wire+Ma, Ac	1 – 40		
Sonya Type 1265	Bulgaria			1x2 – 30, 1x2 – 25		5
Sonya Type 1265	Cuba			1x2 – 30 1x2 – 25		8
Sonya Type 1265	Ethiopia & Eritrea			1x2 – 30 & 1x2 – 25		8
Sonya Type 1265	Syria			1x2 – 30 or 2x1 – 30 & 1x2 –25		8
Sonya Type 1265	Vietnam			2 – 30 & 1x2 – 25		8
Styrso	Sweden	2xDE	AK–90 ac EL–90 magnetic & wire			
Ton	South Africa			1 – 40		
Vanya Type 257D	Bulgaria			1x2 – 30		8
Vanya	Syria			1x2 – 30		8
Vegesack	Turkey			1x2 – 20		

Class	Navy	ROVs	Sweep Gear	Guns[1]	Missiles	Mines[†]
Wosao	China		Wire,Ma,Ac	2x2 – 2		6
Yevgenya Type 1258	Bulgaria			1x2 – 25		
Yevgenya	India			1x2 – 25		
Yevgenya	Syria			1x2 – 25		
Yukto I & II	North Korea			1 – 37 or 1x2 – 25		8
Yurka Type 266	Russia		Wire+M,A	2x2 – 30	2x4 SA–N–5/8	10
Yurka Type 266	Egypt	?ROV		2x2 –30		10
Yurka Type 266	Vietnam			2x2 – 30		10
MINEHUNTERS						
CME	Spain	2 Pluto Plus		1 – 20		
MHC (Swiftships type)	Egypt	Pluto				
Bay	Australia	2 PAP 104				
Circe	France	PAP	None	1 – 20		
Flyvefisken	Denmark	DE	None	1 x 76		?
Frankenthal Type 332	Germany	2 x Pinguin		1 – 40	2x4 Stinger	
Gorya Type 1260	Russia	1 ROV	Wire+M,A	1 – 76 1x6 – 30	2x4 SA–N–5	
Huon (Gaeta)	Australia	2xDE	Oropesa wire Mini–Dyad influence	1 – 30 mm		
Lida Type 1259.2	Russia		2 A, 1 wire	1x6 – 30		
Oksoy	Norway	2 Pluto	None	1 or 2 – 20	1x2 Sadral	
Osprey	USA	SLQ–48	SLQ–53			
Sandown	UK	2 PAP 104		1 – 30		
Sandown	Saudi Arabia	2 PAP 104		1x2 – 30		
Swallow	South Korea	2 Pluto	Wire	1 – 20		
Tripartite	Belgium	2 PAP 104	Mechanical	1 – 20		
Tripartite	France	2 PAP 104	AP–4 + wire	1 – 20		
Tripartite	Indonesia	2 PAP 104	OD3 Oropesa wire F–82 Ma & AS 203 Ac	2x1 – 20		
Tripartite	Netheralands	2 PAP 104	OD 3 wire	1 – 20		
Tripartite	Pakistan	2 PAP 104	Wire + MKR 400 Ac & MRK 960 Ma	1 – 20		
Yevgenya Type 1258	Angola			1x2 – 25		
Yevgenya Type 1258	Cuba					
Yevgenya Type 1258	Russia			1x2 – 25		8 racks

Class	Navy	ROVs	Sweep Gear	Guns[¶]	Missiles	Mines[†]
Yevgenya Type 1258	Vietnam			1x2 -25		
Yevgenya Type 1258	Yemen			1x2 - 25		
MINEHUNTERS/SWEEPERS						
MSC 07	Japan	PAP 104	Dyad influence	1x3 - 20		
MSC 322	Saudi Arabia		Wire + Ma	1 - 20		
MWV 50	Taiwan	2 Pinguin		1 - 40		
Adjutant	Greece			1 - 20		
Avenger	USA	2 SLQ-48	SLQ-37Ma/Ac Oropesa SLQ-38 wire ALQ 166 Ma M/S vehicle			
Bang Rachan	Thailand	2 Pluto	Wire+Ma, Ac	3 - 20		
Gaeta	Italy	1 MIN Mk 2 1 Pluto	Oropesa wire	1x2 -20		
Hatsushima	Japan		S 4	1x3 - 20		
Hunt	UK	2 PAP 104	MS 14 Ma MSSA Mk 1 Ac Mk 8 Oropesa wire	1 - 30 2 - 20		
Kingston	Canada	ROV module	M/S module	1 - 40		
Landsort	Singapore	2 PAP 104	SAM wire	1 - 40		2 rails
Landsort	Sweden	2 x DE	Wire+M, A	1 - 40		
Lerici	Italy	1 MIN 77 1 Pluto	Oropesa wire	1 - 20		
Lerici	Malaysia	2 PAP 104	'O' Mis-4 wire	1 - 40		
Lerici	Nigeria	2 Pluto	'O' Mis 4 wire	1x2 - 30 & 2x1 - 20		
Lindau Type 331	Germany	2xPAP 104		1 - 40		
Lindau Type 351	Germany	None	Wire	1 - 40		
Notec II Type 207M	Poland		Wire,+M,A	1x2 - 23	2 SA-N-5	6-24
River	South Africa	2 PAP 104		1 - 20		
Sonya Type 1265/1265M	Russia			2x60 - 30 or 1x2 - 30 & 1x2 - 25	2x4 SA-N-5	8
Ton	Argentina		Mk 3 wire Mk 2 Ma, AD Mk 3 Ac	1 or 2 - 40		
Uwajima	Japan		S 7	1x3 - 20		
Vanya Type 257D/DM/DT	Russia			1x2 - 30 & 1x2 - 25		8-12
Vukov Klanac (Ton)	Yugoslavia	PAP 104**		2 -20		

Class	Navy	ROVs	Sweep Gear	Guns[¶]	Missiles	Mines[†]
Yaeyama	Japan	S 7	S 8 (SLQ–48)	1x3 – 20		
DRONES						
MSD	Australia					
SAM	Sweden		Ma, Ac			
SAM II	Sweden		Ac,Ma + ?E			
SAV	Denmark					
Futi Type 312	China		Ma, Ac			
Futi Type 312	Pakistan		Ma, Ac			
Ilyusha Type 1253	Russia					
Tanya Type 1300	Russia					
Troika	Germany		M, A			
MISCELLANEOUS						
AN–2 (mine warfare/patrol)	Hungary		Wire			? (ground)
MCM (diving tenders)	France					
MCS (training & support)	Russia			2x2 – 25		
MSA(T) (auxiliary M/S tugs)	Australia		AMASS Wire & influence			
MSA(S) (Brolga)	Australia		Wire+Ma, Ac			
MSA(S) (auxiliary M/S)	Australia					
MSA (auxiliary minesweepers)	Canada		Mk 9 wire			
MSI (Swiftships route survey)	Egypt					
MSR (minesweeping launches)	Greece					
MST/ML (minesweeper support)	Japan					M/L
MST (support ship)	Thailand		M/S	1 – 40 & 2 – 20		
YAG (minehunting tenders)	Turkey			1 – 20		
YDT (diving tenders)	Canada					
YDT (diving tenders)	Canada					
Type 742 (diver support ship)	Germany					
Antares (route survey)	France	None	Wire			
Cosar (support ship/minelayer)	Romania			1 – 57 & 2x2 – 30		200
Ejdern (sonobuoy craft)	Sweden					
Fukue (support ship)	Japan			1x3 – 20		
Hayase Class (MST/ML)	Japan			1x2 – 76 & 2x3 – 20		116
Souya Class (MST/ML)	Japan			1x2 – 76 & 2x3 – 20		460

¶ Machine guns not listed
Depth charges are not listed
$ In some
** In minehunters
DE Double Eagle
M Magnetic sweep
A Acoustic sweep
P Pressure sweep
E Electric influence

TABLE 3:4 Mine Warfare Vessel Designs – Machinery

Class	Navy	Engines	Shafts	Thrusters	Auxiliary	Propellers	Speed	Range	Crew
MINESWEEPERS									
K8 (M/S boat)	Vietnam	2/300	2				18		6
MSB (M/S boat)	Thailand	1/165	1				8		10
MSB 07 (M/S boat)	Japan	2/480	2				11		10
MSC (new minesweepers)	Belgium	2/2,176	2	?	None	fp	15	3,000/12	26
MSC (river minesweepers)	Romania	2/870	2				13		
MSC 268	South Korea	2/880	2				14	2,500/14	40
MSC 268	Pakistan	2/880	2				13.5	3,000/10.5	39
yMSC 268 & 292	Iran	4/696[1]	2				13	2,400/10	40
MSC 289	South Korea	4/696	2				14	2,500/14	40
MSC 294	Greece	2/1,760	2				13	2,500/10	39
M 15	Sweden	2/320	2				12		10
M 301	Yugoslavia	2/?	2				12		
PO 2 Type 501	Bulgaria	1/300	2				12		8
T 43	Albania	2/2,000	2	None			15	3,000/10	65
T 43 (China built)	Bangladesh	2/2,000	2				14	3,000/10	70
T 43 Type 010	China	2/2,000	2				14	3,000/10	70
T 43	Egypt	2/2,000	2				15	3,000/10	65
T 43	Indonesia	2/2,000	2				15	3,000/10	77
T 43 Type 254	Russia	2/2,000	2				15	3,000/10	65
T 43	Syria	2/2,000	2				15	3,000/10	65
T 301	Albania	3/900	3	None			14	2,200/9	25
T 301	Romania	3/900	3	None			14.5	2,200/9	25
Adjutant & MSC 268	Spain	2/880	2				14	2,700/10	39
Adjutant & MSC 268	Taiwan	2/880	2				13	2,500/12	35
Adjutant, MSC 268 & MSC 294	Turkey	4/696	2				14	2,500/10	35
Aggressive	Belgium	4/1,760	2			cp	14	2,400/12	40
Aggressive	Spain	4/2,280	2			cp	14	3,000/10	74
Aggressive	Taiwan	4/2,280	2			cp	14	3,000/10	86
Alta	Norway	2/3,700		2[†]	2/1,740		20.5	1,500/20	32
Arko	Sweden	2/1,360	2				14		25
Baltika Type 1380	Russia	1/300	1			cp	9	1,400/9	10
Bluebird	Denmark	2/880	2				13	3,000/10	31
Bluebird	Thailand	2/880	2				13	2,750/12	43
Cape	Iran	4/1,300	2				13	1,200/12	21
Cove	Turkey	4/696	2				13	900/11	25
Dokkum	Netherlands	2/2,500	2				16	2,500/10	27–36
Frauenlob Type 394	Germany	2/2,200	2				12+	700/14	25
Gassten	Sweden	1/460	1				11		
Gilloga	Sweden	1/380	1				10		

Class	Navy	Engines	Shafts	Thrusters	Auxiliary	Propellers	Speed	Range	Crew
Ham	Yugoslavia	2/1,100	2				14	2,000/9	22
Hameln Type 343	Germany	2/6,140	2			cp	18		37
Hisingen	Sweden	1/380	1				10		
Kiiski	Finland	2/340				2 water jets	11	260/11	4
Kondor II Type 89	Indonesia	2/4,408	2			cp	17	2,000/14	31
Kondor II Type 89.2	Latvia	2/4,408	2			cp	17		31
Kondor II	Uruguay	2/4,408	2			cp	17	2,000/15	31
Krogulec Type 206F	Poland	2/3,750	2				18	2,000/17	48
Kuha	Finland	2/600	1			cp	12		15
Leniwka Type 410S	Poland	1/570	1				11	3,100/8	16
Lienyun	Vietnam	1/400	1				8		
Musca	Romania	2/4,800	2				17		
Natya I Type 266M	Ethiopia & Eritrea	2/5,000	2			cp	16	3,000/12	65
Natya I Type 266M	India	2/5,000	2			cp	16	3,000/12	82
Natya I Type 266ME	Libya	2/5,000	2			cp	16	3,000/12	67
Natya I/II Type 266M/DM	Russia	2/5,000	2			cp	16	3,000/12	67
Natya I Type 266M	Syria	2/5,000	2			cp	16	3,000/12	65
Natya I Type 266ME	Yemen	2/5,000	2			cp	16	3,000/12	67
Nestin (river)	Hungary	2/520	2				15	860/11	17
Nestin (river)	Iraq	2/520	2				12	860/11	17
Nestin (river)	Yugoslavia	2/520	2				15	860/11	17
Notec Type 207P	Poland	2/1,874	2				14	1,100/9	24
Olya Type 1259	Russia	2/471	2				12	500/10	15
Olya Type 1259	Bulgaria	2/471	2				12	300/10	15
River	Bangladesh	2/3,100	2				14	4,500/10	30
River	UK	2/3,100	2				14	4,500/10	30
Schutze	Brazil	4/4,500	2			2 cp	24	710/20	39
Sonya Type 1265	Bulgaria	2/2,000	2				15	1,500/14	43
Sonya Type 1265	Cuba	2/2,000	2				15	3,000/10	43
Sonya Type 1265	Ethiopia & Eritrea	2/2,000	2				15	3,000/10	43
Sonya Type 1265	Syria	2/2,000	2				15	3,000/10	43
Sonya Type 1265	Vietnam	2/2,000	2				15	3,000/10	43
Styrso	Sweden	2/1,104	2	1			13		17
Ton	South Africa	2/3,000	2				15	2,300/13	27
Vanya Type 257D	Bulgaria	1/2,502	1				16	2,400/10	36
Vanya	Syria	1/2,500	1				16	1,400/14	36
Vegesack	Turkey	2/1,500	2			cp	15		33
Wosao	China	4/4,400	4				25	500/15	40
Yevgenya Type 1258	Bulgaria	2/600	2				11	300/10	10
Yevgenya	India	2/600	2				11	300/10	10
Yevgenya	Syria	2/600	2				11	300/10	10
Yukto I	North Korea	2/	2				18		22
Yukto II	North Korea	2/	2				18		22
Yurka Type 266	Russia	2/5,350	2				17	1,500/12	45

Class	Navy	Engines	Shafts	Thrusters	Auxiliary	Propellers	Speed	Range	Crew
Yurka Type 266	Egypt	2/5,350	2				17	1,500/12	45
Yurka Type 266	Vietnam	2/5,350	2				17	1,500/12	45
MINEHUNTERS									
CME	Spain	2/1,523		2	2/150	2 VS	14	2,000/12	40
MHC (Swiftships type)	Egypt	2/1,068		1/300		2 Schottel	12.4	2,000/10	25
Bay	Australia	2/650	2 Schottel				10	1,500/10	14
Circe	France	1/1,800	1				15	3,000/12	48
Flyvefisken	Denmark	1/5450 GT 2/5800	3	1 bow	1/500	cp (outer)	30	2,400/18	19–29
Frankenthal Type 332	Germany	2/5,550	2		1	cp	18		37
Gorya Type 1260	Russia	2/5,000	2				17		70
Huon (Gaeta)	Australia	1/1,986	1		3,506	cp	14	1,600/12	36
Lida Type 1259.2	Russia	3/900	3				12	650/10	14
Oksoy	Norway	2/3,700		2[1]	2/1,740		20.5	1,500/20	38
Osprey	USA	2/1,600		1/180 bow	2/360	2 VS	12	1,500/10	51
Sandown	Saudi Arabia	2/1,500	2	2 bow		VS	13	3,000/12	34
Sandown	UK	2/1,500	2	2 bow		VS	13	3,000/12	34
Swallow	South Korea	2/2,040		1 x bow/102		2 x VS	15	2,000/10	44
Tripartite	Belgium	1/1,860	1	2	2/240	2 x cp	15	3,000/12	46
Tripartite	France	1/1,860	1	2	2/240	2 x cp	15	3,000/12	46
Tripartite	Indonesia	2/2,610	2	2 bow/150	2/240	2 Schottel	15	3,000/12	46
Tripartite	Netherlands	1/1860	1	2	2/240	2 x cp	15	3,000/12	29–42
Tripartite	Pakistan	1/1,860	1	2 bow	2/240	2 x cp	15	3,000/12	46
Yevgenya Type 1258	Angola	2/600	2				11	300/10	10
Yevgenya Type 1258	Cuba	2/600	2				11	300/10	10
Yevgenya Type 1258	Russia	2/600	2				11	300/10	10
Yevgenya Type 1258	Vietnam	2/600	2				11	300/10	10
Yevgenya Type 1258	Yemen	2/600	2				11	300/10	10
MINEHUNTERS/SWEEPERS									
MSC 07	Japan	2/1,800	2				14		40
MSC 322	Saudi Arabia	2/1,200	2				13		39
MWV 50	Taiwan	2/2,180	2				14	3,500/14	45
Adjutant	Greece	2/880	2				14	2,500/10	38
Avenger	USA	4/2,400	2	1/350	2/400	cp	13.5		81
Bang Rachan	Thailand	2/3,120	2		1 motor	cp	17	3,100/12	33
Gaeta	Italy	1/1,985	1	3 x 506	3/1,481	cp	14	1,500/14	47
Hatsushima	Japan	2/1,440	2				14		45
Hunt	UK	2/1,900	2	1 bow	1/780		15	1,500/12	45
Kingston	Canada	4–2/3,000[4]		2 Z drive			15	5,000/8	37
Landsort	Singapore	4/1,592				2 VS	15	2,000/12	38
Landsort	Sweden	4/1,592				2 VS	15	2,000/12	29
Lerici	Italy	1/1,985	1	3 x 506	3/1,481	cp	14	1,500/14	47
Lerici	Malaysia	2/2,605	2		3/1,481	cp	16	2,000/12	42
Lerici	Nigeria	2/3,120				2 water–jets	15.5	2,500/12	50

Class	Navy	Engines	Shafts	Thrusters	Auxiliary	Propellers	Speed	Range	Crew
Lindau Type 331	Germany	2/4,000	2			2	16.5	850/16.5	43
Lindau Type 351	Germany	2/5,000	2			2	16.5	850/16.5	44
Notec II Type 207M	Poland	2/1,874	2		2/816		13	790/14	24
River	South Africa	2/4,515				2 VS	16	2,000/13	40
Sonya Type 1265/1265M	Russia	2/2,000	2				15	3,000/10	43
Ton	Argentina	2/3,000	2				15	2,500/12	27–36
Uwajima	Japan	2/1,400	2				14		40
Vanya Type 257D/DM/DT	Russia	1/2,502	1				16	1,400/14	36
Vukov Klanac (Ton)	Yugoslavia	2/1,620	2				15	3,000/10	40
Yaeyama	Japan	2/2,400	2	1 bow/350			14		60
DRONES									
MSD	Australia	2/300					45		
SAM	Sweden	1/210				1 Schottel	8	330/8	
SAV	Denmark	1/350				1*	12		4
Futi Type 312	China	1/300				cp	12	144/12	3
Futi Type 312	Pakistan	1/300				cp	12	144/12	3
Ilyusha Type 1253	Russia	2/500	2				12	300/10	10
Tanya Type 1300	Russia	1/270	1				10		
Troika	Germany	1/446	1				10	520/9	3
MISCELLANEOUS									
AN–2 (mine warfare/patrol)	Hungary	2/220	2				9		6
MCM (diving tenders)	France	2/2,200	2	1/70			13.7	2,800/13	14
MCS (training & support)	Russia	1/2,176	1				15	5,400/12	38
MSA(T) (auxiliary M/S tugs)	Australia	2/2,400	2				11	6,300/10	10
MSA(S) (Brolga)	Australia	1/540	1			cp	10.5		8
MSA(S) (auxiliary M/S)	Australia	1/480–359	1				10		8
MSA (auxiliary minesweepers)	Canada	4/4,600	2	1 bow/575			13	12,000/13	41
MSI (Swiftships route survey)	Egypt	2/928	2	1/60			12	1,500/10	16
MSR (minesweeping launches)	Greece	1/60	1				8		6
MST/ML (minesweeper support)	Japan	2/19,800	2				22		160
MST (support ship)	Thailand	2/1,310	2				12		77
YAG (minehunting tenders)	Turkey	2/2,000	2				20		
YDT (diving tenders)	Canada	2/165	2				11		9
YDT (diving tenders)	Canada	2/228	2				11		13
Type 742 (diver support ship)	Germany	2/5,200	2				19		11
Antares (route survey)	France	1/800	1	1		cp	10	3,600/10	25
Cosar (support ship/minelayer)	Romania	2/6,400	2				19		75
Ejdern (sonobuoy craft)	Sweden	2/366	2				15		9
Fukue (support ship)	Japan	2/1,440	2				14		38
Hayase Class (MST/ML)	Japan	4/6,400	2				18		180
Souya Class (MST/ML)	Japan	4/6,400	2				18	7,500/14	185

* Pump jet propulsion
¶ Water jets

VS Voith Schneider propellers
GT Gas Turbine
§ Except MSC 268 – 2/880
μ Diesel electric drive, 4 diesels, 2 motors

The Troika control ship Paderborn of the German Lindau class *(H M Steele)*

SECTION 4 - MINE WARFARE VESSEL COMMAND & TRACKING SYSTEM INVENTORIES

4.1 Introduction

Great advances have been made in data handling and the provision of command and control capabilities for mine countermeasures vessels. These result primarily from major developments in the field of electronics. The latest generation of command systems allow various mine warfare systems to be more effectively deployed. Improvements in sonar systems and other modern sensors provide the command system with a much higher volume of more accurate data than was previously available. Integrating and processing all this data enables the entire region around the mine warfare vessel to be accurately surveyed, and for targets to be detected, classified and neutralised with a much higher degree of probability than was previously possible.

With the large volume of data now being made available to the command, a co-ordinated, effective means of correlating and presenting this wealth of information is essential if the MCMV is to carry out its task effectively and within a reasonable time span. Modern tactical plan displays using colour graphics indicate to the command the route to be swept or surveyed, the ship's position, all objects on the seabed, the format of the seabed with integrated contour lines, the direction of the sonar beam, danger and specified circles around objects, map overlay, and alphanumeric tote display. As they are of modular construction and use distributed processing and pre-processing techniques, these systems can integrate with a wide variety of minehunting sonars and other MCM systems now available allowing the command to accept only valid data required for the task in hand.

This huge bank of data has to be assimilated and processed at high speed so that an accurate picture of the complete underwater environment can be generated. It requires programmable, high capacity computers which have the ability to simultaneously retrieve and handle vast amounts of data from a wide variety of sources, process that data and present the command team with an accurate tactical picture of the area under surveillance from which they can make effective decisions.

The most efficient way of handling data is via the database. The ability to have all data passing round a ring system has enabled the multipurpose console to be developed. These are normally allocated to a specific task, but can be switched to undertake a different function depending on the priority and nature of the requirement. This multi-role capability provides enormous flexibility and redundancy.

Among features now being incorporated into the latest command systems are window techniques which allow the display of information such as raw sonar data, contact data and track data and the possibility of correlating and extracting radar position data, ESM data and so on. Other features under study include the capability to automatically control the ship in various attitudes such as hovering and heading using autopilot and propulsion control. The problem with propulsion control is integrating the thrust developed by the various manoeuvring control systems. Development of such an automatic control system will greatly ease the strain on the crew during minehunting operations.

Finally, the man/machine interface (MMI) has to be of the highest order. The presentation of the vast amounts of data required in minehunting demand displays of high resolution and clear definition, the careful use of colour techniques, and the possibility of presenting more specific data from various sensors using window techniques.

The key to MCM is accurate navigation and precise positioning. It is often necessary for vessels to return to the position of previously defined contacts at a later date and to re-examine them or, in the case of confirmed mines, to carry out countermining. The need for extremely precise positioning and plotting, particularly of objects on the seabed is absolutely vital if an MCMV is to carry out this task effectively. With such a system a vessel can return to the precise position where a contact was previously found with considerable saving in time and with the sure knowledge that the original contact will be found again.

The most effective form of MCM is that of route surveying. In peacetime this is a primary task not only of the MCM force but also of the hydrographic service. To be effective, route surveying requires that the whole of a proposed wartime shipping route be surveyed extremely carefully to produce bottom contour charts which show in the minutest detail the composition of the seabed and precise data on all objects on it. By regularly surveying and updating the charts, accurate pictures of proposed routes can be maintained which will enable safe ones to be selected in wartime. Furthermore, these routes can be checked by the minehunters much more rapidly and accurately as the vessel then only has to check objects not already marked on the chart.

All these developments have had a major impact on the design and capability of modern mine warfare combat information systems and on the design of mine countermeasures vessels themselves. It is now possible to achieve considerable reduction in the size and weight of equipments. This in turn will lead to a more effective allocation of space within the already constricted volume available aboard a modern mine countermeasures vessel. The need to provide space for an ever increasing array of mine warfare equipment (wider range of sensors, ROVs, neutralising systems, handling equipment and so on) is placing an increasing burden upon ship designers, and any developments which result in more space being made available within a limited volume is of considerable benefit.

4.2 Europe & Scandinavia

FRANCE

Evec 20

The Evec 20 mine warfare data handling system was developed for the *'Tripartite'*. Based on a horizontal plotting table and screen on which up to 200 sonar contacts can be displayed the system receives data from the central processing unit which has a 20 kbyte random access memory. It samples and processes data from the sonar, radio navigation systems, navigation radar, gyrocompass and doppler log to present the operator with a continuously updated display of the minehunting area showing position of own ship, located targets, radar contacts, and tracks and lines representing specified zones. All or part of the displayed images can be recorded on cassette for subsequent recall, updating and maintaining an up-to-date permanent record of operations relating to a specified zone. A repeater on the bridge displays point co-ordinates from the minehunting sonar. The computer in the EVEC 20 also assists the automatic pilot, providing command inputs and error generation, and helps to control the automatic radar navigation systems providing location computations, error corrections, location of tracking windows and so on. The automatic pilot is used to keep the ship on a predetermined track during minehunting operations, receiving necessary data from the gyrocompass and doppler log. The navigation radar and automatic tracking system integrate with the Evec 20 to provide a buoy location system.

Ibis Systems

The Ibis is a family of mine countermeasures data handling systems. Ibis III integrates the Thomson TSM 2021B (DUBM 21B) dual–antenna sonar with a Thomson TSM 2060 Naviplot tactical plotter for navigation, plotting the precise location of detected targets and the recording of all relevant data. Data are displayed on a large, four–colour CRT which is mounted at an angle to enable the operator to work in a sitting position, giving him greater safety in the event of a mine exploding close to the ship. For operations in the vicinity of complex geographical formations (such as archipelagos and fjords) the raw radar video can be overlaid on the graphic symbology. In all cases coastlines are presented in the synthetic mode. The Ibis III system provides a continuous simultaneous display of the planned and ship's actual track, fully labelled display of up to 256 contacts, complete recording of all operational data and storage in the computer's memory. This enables a continuous comparison to be made between previously recorded data and the current situation with regard to the area being surveyed. This enables new underwater contacts to be instantaneously identified and located, providing increased safety for the ship and resulting in considerable saving in time on minehunting operations. Associated subsystems of Ibis III are: ship's doppler sonar log (Thomson–CSF 5730); radio navigation system; navigation radar; compass; log; SATNAV; echo–sounder; radio datalink; and Sweepnav, an acoustic location system for mechanical sweeps. As optional extras the Ibis III can be integrated with the ship's autopilot for automatic track following course to a point and hovering. The Ibis III Mk II is an upgraded version of Ibis III which comprises the TSM 2021D (DUBM 21D) sonar and the TSM 2061 tactical system. The Ibis V is a lightweight system suitable for retrofit as well as new construction. The sonar associated with the Ibis V is the Thomson TSM 2022 single array system, which takes up much less space in the hull than the TSM 2021. Whereas the 2021 can provide simultaneous detection and classification in different directions, the 2022 performs detection and classification in sequence and requires only a single operator. The TSM 2026 Naviplot is the other main component of the IBIS V. IBIS V weighs just 1.5 tonnes, the array having a span of about 1.5 m but pivoting to retract vertically into the sonar well, which is only 75 cm in diameter. Ibis V Mk II is an upgraded version which comprises the TSM 2022 Mk II sonar and the TSM 2061 tactical system. The latest system to be developed in the Ibis series is the Ibis 43. Using sophisticated computing equipment and the TSM 2054 side scan sonar, with antenna incorporated in a towfish, Ibis 43 produces high quality, high resolution images of the seabed enabling operators to identify underwater mines. The computer system ensures real–time optimisation of sonar operating parameters depending on the speed of the towed body, its height above the seabed and the automatic compensation for roll, pitch and yaw. Computerised aids are used in conjunction with an image management system to ensure real–time detection and classification of unknown objects and to compare images obtained in successive missions in a given sector. The towfish can be navigated manually or automatically at constant depth between 6 m and 200 m with an altitude in relation to the seabed from 4 m to 15 m, programmable from the sonar console keyboard. It can automatically follow seabed terrains with gradients of more than 15 per cent with high precision. It also has an obstacle avoidance capability for obstacles up to 10 m high at maximum speed.

GERMANY

MWS 80

The MWS 80 minehunting weapons system has been developed for the *'Frankenthal'* class minehunters. The integrated multirole MCM system performs: search; detection; classification; identification and neutralisation of ground and moored mines; precise navigation and vessel control; full control and co-ordination of all MCM operations; and maintains an accurate record of all areas searched, targets located and their geographic position and details of classification results. The MWS 80 integrates the DSQS-11M minehunting sonar, the NBD (navigation and vessel control equipment), the TCD tactical command system for co-ordinating and documenting MCM operations and the mine disposal system, and the AIS 11 active identification sonar installed on the ROV. The DSQS-11M provides independent and simultaneous detection and classification modes through 360° azimuth. The stabilised sonar beams simultaneously cover a 90° horizontal sector and a 60° vertical sector for 3-D target location. Functions include performance prediction, computer-aided detection classification and tracking with full colour display to improve detection, performance and discrimination between targets. The MMI features function keys and interactive, colour-coded control displays. The TCD assists the operator in controlling all phases of an MCM operation by providing means for display, plotting, storage and retrieval of tactical data. The equipment employs a comprehensive database which is continuously updated during operation. The navigation sensors such as GPS, radio location systems, gyro or inertial platform and so on, and the integrated doppler Log DLO 3-2 enable the NBD to compute with a very high degree of accuracy the various navigational parameters such as ship's geographical position, groundspeed, speed through water, course, heading and drift. All data handled are recorded on hard disk, tape and printer to enable missions to be stopped and started at will, provide recall of data for comparison and evaluation, maintenance of a permanent record of the seabed for MCM survey and for training purposes ashore.

TCD (Tactical Command and Documentation Equipment)

The TCD forms an integral part of the MWS 80 minehunting weapons system. It provides comprehensive facilities allowing the preparation, control and documentation of all phases of the MCM operation. A variety of functions is available on the two consoles including: track planning for the selected search area, taking into account tactical considerations such as sonar performance data and environmental conditions; presentation of tactical, navigation and status information, including radar video and live sonar displays; presentation of manoeuvre data to the helmsman; control of sonar and navigation equipment; contact management with correlation of detected and stored sonar targets; documentation of all results for evaluation and follow-on missions; and data logging of selected parameters. Virtually all these functions are linked by the TCD's comprehensive tactical database management system. The consoles incorporate two high-resolution raster scan displays and keypads for direct system operation. Peripheral equipment includes a mass memory unit, printer, plotter, manoeuvre display and the minehunting sight.

NVC (Navigation and Vessel Control Equipment)

The NVC integrated navigation and vessel control equipment system forms part of the MWS 80 minehunting weapons system. It combines the functions of precise navigation, vessel control and precise speed measurement, performed by an integrated doppler log. Data is accepted from a wide range of navigational sensors such as GPS, radio location system, Decca, inertial platform, gyro, echo-sounder, anemometer and radar. From these inputs the NVC computes optimal values for position, heading, course, longitudinal and transverse speed over ground and through the water, set drift and water depth by Kalman filtering. By providing datum transformations the NVC will accept and output navigational data in most of the commonly used reference systems. Using track and waypoint data from the tactical database or entered directly at the control and display unit, the NVC automatically steers the vessel along the planned search pattern. A manual vessel control mode via joystick is available.

ITALY

Mactis MM/SSN-714(V)2

The Mactis MM/SSN-714(V)2 minehunting digital navigation and plotting data processing system was developed for the Italian Navy's *'Gaeta'* class mine countermeasures vessels. The main functions of the system are: operations planning; automatic computation and presentation of the ship's current position; navigation control (through autopilot interface); display of the tactical situation; location of surface and underwater targets; analysis and presentation of target characteristics; guidance of surface and underwater craft; and event recording. The system's computer interfaces with recording units, display units, controls, printer/plotter and ship's sensors including: radar; sonar; compass; log; and various other navigation aids. The operator's display has a vertical 16 in graphic video screen, an alphanumeric display, a keyboard for communications with the system, supplementary data readout display units and associated input controls on the bridge and in the operations room. The system integrates with the SQQ-14 mine-detection and classification sonar for the analysis and presentation of target characteristics. Comprehensive navigation equipment, in addition to the MM/SPN-703 navigation radar, is provided to ensure precise positioning of the ship and accurate plotting of underwater objects encountered.

NORWAY

Micos

The Micos minehunting command and control system has been developed for the Norwegian mine countermeasures programme. Micos is an integrated mission control system which can be configured for either minehunting or minesweeping. The configuration on board the *'Oskoy'* class minehunters comprises: an integrated bridge system; an MCM command and control system; the sonar system; and ROV and mine disposal system. The configuration on board the *'Alta'* class minesweepers comprises: an integrated bridge system; an MCM command and control system in the operations room; the sonar system; interfaces to the sweep gear. The integrated bridge system comprises a Simrad ADP 701 navigation and dynamic positioning system for high-precision navigation and automatic control of the vessel, and a Norcontrol DB 2000 officer-of-the watch ARPA system for surface surveillance. The MCM command and control system comprises two Simrad ATC 900 multifunction tactical consoles, two operator consoles for the Simrad/Thomson Marconi Sonar Ltd TSM 2023 detection and classification minehunting sonars and a control console for the ROV operator. The system components are all integrated via a dual-redundant Ethernet LAN. Micos is based on an open architecture design using a high degree of COTS. The command and control system has a wide range of functions for planning, operation and reporting and documentation of an MCM mission. It comprises a database containing 2D electronic charts, 3D bottom terrain models, previous observations of bottom objects, navigation routes and so on. Also available are functions for performing MCM calculations for coverage and system performance analysis, including updating of the MCM database. The ADP 701 dynamic positioning system includes functions for automatic control of the vessel comprising a full three-axis manual mode, autopilot (for transit), hovering mode, auto-track and auto-sail modes. The two latter modes are used to steer the vessel automatically on a planned course with minimum cross-track error. The system interfaces with the sweep gear for automatic calculation of the equipment's position and coverage and for measuring and counteracting the forces induced by the gear on the vessel.

Satyr

The Satyr minefield inspection, control and maintenance command and control system is an integrated mission control system adapted from the Micos command and control system. It comprises: a Simrad tactical command and control system; a Simrad ADP 701 dynamic positioning system; a Simrad HPR 410 hydro–acoustic reference system; a Simrad EA 300 echo–sounder; the Artemis surface position reference system; and a windspeed/direction sensor. The system components are all integrated via a dual–redundant Ethernet LAN. Satyr is based on an open architecture design using a high degree of COTS. The command and control system has a wide range of functions for planning, operation and reporting and documentation of missions for minefield laying, inspection and maintenance. The system comprises a database containing 2D electronic charts and high–resolution 3D bottom terrain models. The ADP 701 system installed in the *Tyr* includes in addition to the standard dynamic positioning functions, a 'Follow ROV' mode where the vessel automatically follows an ROV, maintaining an operator–defined distance between the ROV and the vessel.

HPR–300 Positioning System

The HPR–300 portable hydro–acoustic positioning reference system consists of a small transceiver unit in a portable, splashproof cabinet with connectors for transducer, display, joystick, gyro and data output. Normally the system is delivered with mini transponders specifically designed to give position to small ROVs. Using a standard RGB monitor the control of functions and parameters is carried out by joystick. Operating instructions are built into the system and this appears on a self–explanatory display menu. The HPR–300 system has a built–in roll and pitch sensor in the transducer which enables easy and quick installation. The system is able to position five transponders simultaneously, and can be upgraded to position 14 transponders. The transducer may be mounted over the side of a craft of opportunity.

ADP 70X Family

The ADP 70X family of automatic ship control and dynamic positioning systems comprises the ADP 703 triple–redundant system; the ADP 702 dual–redundant system; and the ADP 701 single computer system. The systems can be interfaced to a great number of different navigation systems and can be adopted easily to different thruster/propulsion system configurations. The ADP 701 system includes functions for automatic control of the vessel comprising full three–axis manual mode, autopilot (for transit), hovering mode, and auto–track and auto–sail modes for track–keeping. The two latter modes are used to keep the vessel automatically on a planned track with minimum cross–track error. The system installed on board the Norwegian MCMVs includes a set of functions especially adopted for minehunting and minesweeping operations, for example, 'circle a point', where the vessel is automatically steered around a defined geographical point on the bottom to allow the sonar operator to observe the bottom object of interest at different angles. A 'Follow ROV' mode is also available where the vessel automatically follows an ROV maintaining an operator–defined distance between the ROV and the vessel.

SWEDEN

9 MJ 400

Developed as a joint project between CelsiusTech and Racal (UK) the 9 MJ 400 integrated navigation and combat information system provides a wide range of navigation and MCM functions together with sensor interfaces. The 9 MJ 400 enables the operator to plan in minute detail all manner of MCM tasks. It computes and displays the navigation plan to and from a search area, and during the MCM task displays computerised search tracks, taking into account the type of sonar fitted and environmental conditions. The sonar search is planned using two types of search plan displayed on the conference type combat information console. The operator then defines an area or route to be searched and enters the anticipated sonar coverage and required percentage overlap. The system then computes and displays a search plan, the sonar being controlled in one of two remote-control modes (search or target indication). The search plan and actual tracks, together with the position of located targets, are automatically drawn on the X-Y plotter. During a search the operator can track other ships and aircraft operating in the area, the system tracking 16 targets automatically or 30 targets semi-automatically. Data on mine contacts are stored in the direct access memory of the computer (up to 100) and 'dumped' on to magnetic tape, as well as being fed to the plotter and/or the printer. The magnetic tape can store data on 1,000 mine contacts, which can be read into the direct access memory when planning the operation. All data recorded on tape or printer are available as a permanent record for subsequent comparison when carrying out searches of specific channels. Own ship position and true speed are continuously calculated from each sensor input which includes Decca Navigator, transponders, navigation radar and dead reckoning. The operator selects the most reliable source for input to the system. The computer can store up to four range-measuring beacons when using a microwave transponder system. When linked with the CelsiusTech 9 LV 100 fire control system, the 9 MJ 400 offers an upgraded self-defence capability for the ship as well as improving MCM tracking functions. The command system also features an integrated datalink for the onward transmission to shore-based control centres of important MCM data, as well as to other units in the operating area.

UNITED KINGDOM

System 880

The Racal System 880 minesweeping/minehunting control system provides precise navigation for minesweeping operations and co-ordination between ships operating as part of a team. As such it is suitable for minesweepers such as the '*River*' class vessels of the Royal Navy and Bangladesh Navy and is recommended for Craft Of Opportunity (COOP). A minehunting version of the System 880 is also available using a comprehensive tactical display in addition to the equipment available on the minesweeping package. The system can integrate with a variety of sonars and is capable of controlling an ROV. The modular system uses distributed processing and can be reconfigured easily to meet different requirements and to interface with different equipments. System 880 comprises two major subsystems: the Racal Integrated Navigation System (RINS) and Racal Action Display System (RADS). RINS is designed for minesweeping vessels, while for minehunters the addition of RADS provides a full MCM capability. System 880 automatically gathers information from sensors and navigational aids and presents it to the command on a ground-stabilised 19 in colour CRT display. The display uses colour graphics to identify route, ship's position, objects, direction of sonar beam, alphanumeric tote display, and danger and safety circles prescribed around the object under examination, all presented on a plan display. Associated with the main display is a control panel, alphanumeric keyboard and smaller alphanumeric display for online communication with the computer system. RINS provides a minesweeper with mission planning capability, accurate navigation display data for the helmsman to aid accurate track-keeping, hard copy printout on a plotter or automatic chart table, and autopilot control. For team sweeping a special purpose version of RINS is available called RAFTS (Racal Aid For Team Sweeping). This provides all the facilities noted above with, in addition, a communications link between the RINS systems on the ships involved in the team sweep to ensure that the wing ship is automatically maintained on station with respect to the lead ship. RADS can be added to RINS to provide integrated sonar, radar, MCM database and a tactical display for minehunting. RADS also provides an acoustic navigation capability enabling inputs from an underwater acoustic positioning system such as Racal's Aquafix 4 to navigate a ROV.

Mains

The Mains minehunting action information and navigation system provides the following main functions: accurate navigation; ship control guidance; integration of minehunting sonar; combined surface/subsurface tactical display; automatic hard copy tactical plot with detailed data printout; with patrol navigation/combat information operating characteristics as an optional second role. Being of modular construction Mains is capable of interfacing with a wide variety of equipments suitable to meet individual requirements of any navy. Own ship position is fixed using one or more high–accuracy radio and microwave ranging systems such as the Racal Hi–Fix 6 (large area coverage extending to 150 km), Hyperfix, or the portable line of sight Trisponder with a range of 80 km. Other navaids include the statutory navigational radar which has sufficient resolution to track marker buoys, SATNAV, Decca, Loran C, doppler sonar, gyrocompass, and conventional electromagnetic or doppler/acoustic log. The Mains computer can also be used for ship control and steering guidance. An automatic plotter and printer provides a hard copy record of operations necessary for subsequent operations covering the same area. The interactive PPI display is the Racal ED1202 which combines both the radar surface situation and own vessel position relative to sonar search plan together with underwater contacts as a single presentation. The displayed picture has multilevel brilliance and correlated data and is presented in alphanumeric form. The computer software is modular and is capable of accepting a variety of languages.

Nautis–M

The Nautis–M integrated command, control and navigation system for mine countermeasures was originally developed for the Royal Navy's '*Sandown*' class. It is designed to integrate the Type 2093 variable depth sonar, a remotely controlled mine disposal system, multiple navigation and environmental sensors, and radar and ship control functions, as an integrated combat system. Functional requirements for Nautis–M cover all phases of the minehunting operation including mission planning, route surveys, minehunting, classification, disposal and mission reporting. The basis of each Nautis system is a new technology autonomous intelligent console with high–resolution shock–hardened colour raster displays with radar video superimposed from a scan calculator within each console. The console includes interfaces for sensors, weapons and, if required, a datalink, an integral radar auto–tracker, processors and memory to handle an extensive command system database. The high–definition colour raster display presents a labelled radar picture with tactical graphics superimposed. A Nautis system comprises a number of such consoles networked via a dual–redundant digital highway which is used to maintain automatically a replica of the command system database within each console. Each console user has independent access to the database and to interfaced sensors, weapons and peripherals according to his task requirements. User facilities enable tasks to be readily changed to give functional interchangeability between consoles of a system. Should any console be out of use then the other consoles remain fully operational as a system, using duplicated interfaces, and can take up the additional tasks. Nautis–M systems can be configured with one or two networked consoles in a ship's operations room and, if required, with a further console on the bridge (as in the '*Sandown*' class). The number of consoles depends on operational requirements, the performance of the overall combat system and the number of operators required to handle the system data.

The Nautis–M online database covers an area of up to 2,000 square nautical miles. The data include route plans, above and below water geographic and environmental data, known seabed contacts and other supporting data that will enable a mission to be planned and carried out with best use made of the ship and its combat system. Examples of database content are: at least 200 radar and sonar tracks; threat evaluations and weapon assignments; route/search plans and navigation status; six user–designated tactical maps; 15 synthetic charts; 200 labelled reference points; 5,000 fully detailed sonar contacts for MCM; and 32 labelled bearing lines and operational warning messages. Operator facilities include: radar; labelled graphics; totes and an interactive main area clearly displayed in one viewing area; two different display compilations maintained for alternate selection by single–key action; display scales typically from 0.125 nautical miles to the limits of the database; off–centring anywhere within the database area; true/relative motion stabilisation; and labelled electronic range and bearing lines which can be hooked onto fixed points and moving tracks. The man/machine interface also includes a typewriter keyboard, up to 32 assignable special function keys, a trackerball and electroluminescent panel. Operational effectiveness of the ship's systems is maximised by full integration of sonar, navigation, collision avoidance and ship control systems through Nautis–M together with the online presentation of the MCM database. The modular architecture of Nautis–M enables it to be readily adapted for MCM combat system applications with various forms of variable depth and hull–mounted sonars, influence and mechanical sweeps, sweep monitoring equipment, navigation systems, radars, mine neutralisation systems and ship control systems. Nautis–M is in service with the UK Royal Navy, Royal Saudi Navy, is being retrofitted to the US *'Avenger'* class (where it is known as the AN/SYQ–15), and has been selected by the Spanish Navy for its new CME class MCMVs.

QMSS

The QMSS mine surveillance system is a portable system for COOP and STUFT as well as offering enhanced mine surveillance and analysis capability to dedicated MCMVs. The system provides real–time targeting and analysis of data logged during surveys, search and salvage and similar operations. The complete package has been based on easily installed, portable commercial survey systems, and includes a high–resolution sidescan sonar, the Trac integrated navigation/positioning system, and the QASAR sonar analysis and reduction system. Trac is an essential element of the QMSS package and provides repeatability of vessel position, underwater ROV and bottom target, together with absolute co–ordination, error modelling and database facilities. As the sonar sweeps the area, the QASAR system helps the sonar operator to look more closely at the targets, to accurately pinpoint their positions and to make comparisons with previously noted seabed objects and data stored on the Trac database. It provides full real–time integration of sonar and position, high–resolution logging of sonar data to optical disc, and incorporates a range of digital sonar enhancement techniques. Built as a compact, transportable unit, QASAR can link any sidescan sonar to any integrated navigation equipment. Sidescan information is displayed on a high–resolution touch–sensitive display screen to give a continuous concurrent picture of the seabed. The user merely points to any object of interest on the waterfall display and the computer's cursor automatically follows the operator's finger to give an initial position which can then be fine–tuned using the system's trackerball. The operator then simply touches or clicks on an on–screen classification menu to categorise the target and pass the relevant information to the Trac screen. Video enhancement techniques, including contrast stretching, allow the operator to look more closely at the target. QASAR logs the sonar data to optical disc and integrates it with data from the positioning system. Later the operator is able to fast–scan the record to review omissions, freeze, zoom and enlarge targets, overrule earlier interpretations, and add and remove contacts. A complete overview of the mission or isolation of specific track lines and groups of targets can then be plotted.

Trac & Chart Systems
The Trac and Chart integrated navigation, hydrographic survey and data logging systems are a family of automated data acquisition and processing systems are used extensively by for a wide variety of operational tasks including mine warfare. Trac provides integrated navigation, data logging, and precise and dynamic positioning for vessel control. It can form an important part of a naval command system by providing real–time data in a variety of formats directly from the vessel's sensors. Trac is a self–contained system which uses an advanced position computation algorithm to handle up to 20 position lines which may be any combination of hyperbole, ranges, bearings or latitude/longitude lines. The system can operate on any spheroid and projection and takes into account the various geodetic reference systems. Control of Trac is through a keyboard, rollerball and colour display. The system supports a wide range of peripherals, including remote colour displays, plotters and printers.

Chart is a complete data processing system which complements Trac. It may be supplied as a stand–alone unit or networked to Trac. The system design supports multiple Trac and Chart workstations networked together to create an integrated multi–user system. Chart may be used afloat or ashore for data analysis and the production of fair sheets, and contains all the necessary processing facilities to carry out these functions.

OE2059
The OE2059 acoustic tracking system is designed for minehunting applications among others. It comprises three units: a controller; transducer assembly; and the underwater transponder (up to five transponders or responders may be tracked by a single control unit, or alternatively a single acoustic pinger). The transducer is a miniature hydrophone array containing three receiving elements and one transmitting element. The transmitting element generates the interrogation pulse to which the transponder replies. The three receiver elements are arranged in an orthogonal pattern, and the transponder return signal is computed from the phase difference between the three elements. In addition to the hydrophone array the transducer incorporates a replaceable plug–in circuit card. The underwater transponder unit generates the acoustic signal which the system tracks. One transponder must be fitted to each underwater unit that requires to be tracked. The transponder generates a 30 kHz acoustic pulse in response to the acoustic interrogation from the transducer.

4.3 Asia, Pacific & Australasia

AUSTRALIA

ASK 4000 Series
The ASK 4000 series of dynamic positioning systems has been designed to provide manoeuvrability and control of all types of vessel. In the most basic system control is exercised using a simple joystick while the most complex system is a triple redundant system. Each system can interface with all known positioning sensors. Control is achieved using clearly identified function keys and comprehensive graphic displays. Expanding systems is achieved by adding one or more modular hardware units which are easy to install. The basic design of each system comprises a standard modular hardware unit which simplifies procurement and eliminates the need for lengthy specification and custom engineering. The hardware requirement is much reduced as the console unit contains all the system elements. The multifunction console is housed in a single cabinet and features a large, ruggedised 20 in

diagonal high–resolution, full colour display with IBM VGA compatibility. Signal processing is carried out using an industry standard processor – a ruggedised VME bus–compatible IBM 80486 digital processor running at high clock frequency. The processing software control algorithms are programmed in high–level language C. Signal modules communicate with digital processors over a high–speed RS–485 data highway. Station–keeping performance data is always shown by using a split screen display. The selected graphics page is shown on the left side of the screen while the permanent display of station–keeping is always shown on the right. Ten information displays are accessed through dedicated panel control keys. The summary page provides a numeric summary of the control process, with heading and position waypoints and actual positions, environmental data such as wind and sea current and a review of the control system values. The sensor page displays individual measurements reported from each sensor. Raw input data is also available for environment and velocity sensors. The alarm page provides full colour coded descriptive alarm messages to indicate alarm conditions and current alarm status. The plot page provides a graphic 'strip chart' display of up to four internal parameters simultaneously. The operator may select any variable for recording together with an appropriate display scale and bias.

4.4 North America, South America & Caribbean

UNITED STATES

PINS

At the heart of the PINS AN/SSN–2(V) (Precise Integrated Navigation System) is a powerful military computer interfaced with an array of commercial and military sensors, control units and data display and recording systems. The PINS uses a Kalman Filter routine to smooth received data and apply appropriate corrections for known error factors. The system integrates and compares data from a wide variety of navigation aids and other sensors to perform precise navigation and position determination, mission planning, plotting and data recording, target location and positioning and post–mission analysis. The Transit SATNAV system is used to provide navigation fixes, interfaced via PINS with the ship's doppler sonar operating in either groundspeed or water speed modes. Other interfaced systems include Loran C and Hyperfix. All contacts and their positions are recorded and displayed, together with the ship's track, on a vertical plotter in the operations room. The high–speed vertical belt–fed plotter provides a hard copy printout using standard navy charts if required.

TABLE 4.1 Command & Tracking Systems

System	Manufacturer	Country	Class Installed on	Navy	Units*
9MJ400	CelsiusTech Systems	Sweden	Landsort	Sweden	7
ADP 701	Simrad Albatross	Norway	Oksoy	Norway	4
ADP 701	Simrad Albatross	Norway	Alta	Norway	5
ASK 4000	Nautronix	Australia	Shoalwater	Australia	1
HPR 300	Simrad Subsea	Norway	?		
MWS 80	STN ATLAS Elektronik	Germany	Frankenthal	Germany	12
MWS 80	STN ATLAS Elektronik	Germany	Bang Rachan	Thailand	2
MWS 80	STN ATLAS Elektronik	Germany	Bay	Australia	2
NVC	STN ATLAS Elektronik	Germany	Frankenthal	Germany	12
OE2059	Simrad	UK	?	?	
PINS	Magnavox	USA	Avenger	USA	14
QMSS	QUBIT Kelvin Hughes	UK			
TCD	STN ATLAS Elektronik	Germany	Frankenthal	Germany	12
System 880	Racal	UK	River	Bangladesh	4
System 880	Racal	UK	River	UK	5
Chart	QUBIT Kelvin Hughes	UK			
Evec 20	Thomson-CSF	France	Tripartite	France	10
Evec 20	Thomson-CSF	France	Tripartite	Pakistan	3
Ibis V	Thomson Marconi Sonar	France	Tripartite	Indonesia	2
Ibis V	Thomson Marconi Sonar	France	Lerici	Malaysia	4
Ibis V	Thomson Marconi Sonar	France	Lerici	Nigeria	2
Ibis V	Thomson Marconi Sonar	France	Landsort	Sweden	7
Ibis V	Thomson Marconi Sonar	France	Landsort	Singapore	4
Ibis V	Thomson Marconi Sonar	France	MWV 50	Taiwan	4
Ibis 43	Thomson Marconi Sonar	France	Flyvefisken	Denmark	4
Mactis	Datamat	Italy	Gaeta	Italy	8
Mains	Racal	UK	Landsort	Sweden	7
Mains	Racal	UK	Swallow	South Korea	6
Micos	Simrad Subsea	Norway	Oksoy	Norway	4
Micos	Simrad Subsea	Norway	Alta	Norway	5
Nautis-M	GEC-Marconi S3I	UK	Sandown	Saudi Arabia	3
Nautis-M	GEC-Marconi S3I	UK	CME	Spain	4
Nautis-M	GEC-Marconi S3I	UK	Sandown	UK	12
Nautis-M	GEC-Marconi S3I	UK	Avenger	USA	14
Nautis-M	GEC-Marconi S3I	UK	Huon	Australia	6
Nautis-M	GEC-Marconi S3I	UK	MSC-07	Japan	2
Satyr	Simrad Subsea	Norway	Tyr	Norway	1
Trac	QUBIT Kelvin Hughes	UK			

* In service or on order

4.5 Market Prospects

The market for mine countermeasures integrated command and tracking systems is to a very large degree dictated by the scale of new construction. However, first generation systems are now reaching an age where they will soon either have to be replaced, or undergo a major upgrading in order for the ships in which they are installed to remain effective in the face of modern mine technology. New systems will be able to handle the greatly increased volume of data which is now being gathered by the latest generation of sensors in order to present the command with a detailed comprehensive picture of the mine warfare scenario.

The refurbishment and upgrading relating to systems already in service, and to a lesser degree installation of systems in vessels which have not previously been fitted with a command system is, however, a fairly limited market. In the future the demand for command systems which can be fitted into smaller mine countermeasures vessels is likely to increase.

The requirements now are for systems which are totally integrated, combining and processing data from a wide variety of sensors dedicated to the detection and classification of mine-like objects, together with systems which can measure all aspects of the underwater environment and combining information from precise navigation equipments and integrating this with a platform management system controlling the vessel's machinery.

Continuing efforts will be made to reduce reliance on human operators, for minehunting can be an extremely tedious task and, when operators become bored or tired, errors can be made that prove fatal. Greater use will be made of datalinks, integrated with the command system, to relay information between vessels on task and the shore-based MCM headquarters. In addition there will be increased integration with the command system of sonar prediction techniques, which will be used as an important means of improving effectiveness and safety with the ability to define safety circles more carefully.

Traditionally, command systems for mine countermeasures vessels have, as with other naval vessels, been proposed by the main shipbuilder, who in the past has always acted as prime contractor. That situation is beginning to change. As electronic systems, and in particular command systems, take on the major function within the vessel, there will probably be a move towards appointing the command system contractor to act as the prime contractor.

Not surprisingly the main national contender in the market for command and tracking systems has, until recently, been France, with a total of 42 systems sold, 32 of them for installation in the vessels of eight overseas navies. Of the systems sold only about a third (15) were for installation in French-built mine countermeasures vessels. The remaining 27 units have been installed in six Italian (*'Lerici'*), 11 Swedish (*'Landsort'*), four Taiwanese (MWV 50) and four models for the Danish *'Flyvefisken'* class vessels.

However, France's traditional lead in the area of mine countermeasures command systems has faced a major challenge from Britain since the start of the decade, with the latest generation system, the GEC-Marconi Nautis having now been ordered by five overseas countries in addition to units sold to the UK Royal Navy. To date a total of 41 Nautis-M systems are operational or on order, of which 29 are for export. There is a vast potential market in Japan for the Nautis-M, and two systems have recently been sold to Japan for installation in the new MSC-07 coastal minehunters currently under construction. The Nautis-M has now captured a major share of the world market, and with both Japan and the United States having bought the system, as well as the systems installed in minehunters of the UK Royal Navy, it is likely that it will become the preferred system for many navies. To what degree the recent tie up between the French Thomson Sintra Company and the sonar arm of the GEC company, both major world contenders in the mine warfare field, will affect the situation remains to be seen.

While France and Britain between them have achieved world domination in command systems for mine countermeasures vessels, they are not the only countries to have developed such systems. Germany and Norway are the other two major contenders in this area of expertise. To date Norwegian systems have only been installed on nine units of the national navy. Germany, however, is also a major force in the development of command systems, but to date has not achieved the same high level of export sales as Britain and France. So far four systems have been sold overseas to two foreign navies.

With France and the UK having secured the major share of the world market, and pursuing an extremely vigorous marketing strategy in this area, it seems unlikely that either Germany or Norway will manage to make major inroads into this field for some years to come.

SECTION 5 - SONAR SYSTEM INVENTORIES

5.1 Introduction

Many of the problems associated with sound transmission in ASW apply equally to MCM, although there are considerable differences between the sonars used in the two forms of warfare. While ASW is essentially a long-range detection problem requiring low frequency, MCM is a very short-range problem requiring high frequencies.

Being primarily a shallow water operation, MCM sonar systems have to overcome problems associated with environmental conditions such as noise, bottom, surface and volume reverberations, the attenuation experienced by sound waves as a result of absorption by micro-organisms, and diffusion and reflection by gas bubbles experienced near the surface, which in shallow water may be excessive when there is high wind, and the heterogeneity of the sea and reflection of sound waves experienced near the surface, at boundary layers and at the bottom. The use of acoustics in minehunting is further compounded by the need to penetrate the top soft layer of a sea bottom to detect buried mines.

The reason for using high frequencies in MCM is because of the need to detect very small objects of varying shape and reflectivity indices, and to produce very high-definition pictures of them. Low-frequency sonar offers extended range but poor definition in detail, while high-frequency gives short range but very high definition. The approach to defining requirements for MCM sonars is therefore quite different from that of ASW, and very specific. This is now becoming even more important as sonars for MCM are required to pick out small mines on the bottom which may be hidden among rocks, buried in mud or sand, protected by anechoic coatings or provided with irregular shapes to reduce the probability of detection and classification. As mines become more stealthy the use of low frequency may become more important as these signals stand a better chance of penetrating anechoic coatings which are being increasingly used on mine casings.

To provide as much detail and data over a very small insonified area as possible requires a high-frequency sonar. The higher the frequency, the greater the volume of data that can be returned to the signal processor, and hence the more detailed the picture derived and presented to the operator. However, the problem with high-frequency sound in water is that it is rapidly attenuated, and so sonar ranges in MCM are very limited. This means either that an MCMV with an onboard sonar probably has to approach within the mine's danger zone in order to detect, identify and classify it, with the possibility of detonating it, or that the sonar must be displaced from the MCMV in some way so allowing the MCMV to remain well outside the danger radius of any possible mine detonation. In the past the use of hull-mounted minehunting sonars has required a trade-off between the high frequency required for high definition, and the lower frequency required to ensure that the MCMV can stand off at a safe distance for any possible mines (the danger radius of ground mines is considered on average to be about 200 m). To overcome this problem most MCM sonars incorporate a dual-frequency capability, using a lower frequency to survey a specific area, before the ship moves in closer to use the sonar's higher frequency with its greater definition capabilities to classify and identify the target. These sonars can either be deployed as hull-mounted sonars or as offboard sonars. Offboard minehunting sonars can be deployed in one of two ways. One way is to use a side scan sonar towed astern of the platform. Alternatively a variable depth sonar can be deployed from a well in the hull of the ship or over the stern

Signal processing has made great strides in recent years and there has been a significant increase in automated detection, classification, tracking and so on and the presentation of data in synthetic form. Modern microprocessors offer significant amounts of processing power available to cope with future demands, without affecting current processing requirements. The problem is knowing how to handle the vast amounts of information available now and in the future to ensure that it is processed in the best way possible to aid the command.

To ensure ship survivability against modern mines, most MCMVs now carry an ROV as well. These vehicles carry either a camera or small high–definition sonar (or both). As a last result in difficult situations a diver may be used to positively identify and neutralise a detected target. The latest generation of ROVs, now the subject of intense development and investigation, include autonomous offboard ahead swimming vehicles capable of carrying out a fully integrated minehunting operation against both shallow and deep water laid mines.

One of the most important factors in mine warfare, as already noted previously, is route surveillance. This requires that areas susceptible to mining must be constantly surveyed well in advance of hostilities. Detailed records of the seabed and all objects on it must be maintained by constantly surveying the area which enables routes to be quickly checked for new targets in time of rising tension and hostilities. Using such records enable routes to be cleared more quickly, or alternative routes to be assigned. Route surveillance requires a high speed of advance (up to 15 knots) and high coverage rate (swath width of about 2,000 m). To achieve these types of coverage requires a sonar range of about 1,600 m. In good propagation conditions a sonar with an azimuth resolution of 3° and a range resolution of 20 nautical miles would be sufficient to carry out such a route survey.

Classification, on the other hand, demands a much slower speed of advance with much greater resolution. Thus speeds of 5 knots are the norm, and to achieve good classification results using acoustic shadow and echo to allow evaluation of the shapes of detected objects at a range of at least 275 m (to ensure platform safety), requires a bearing resolution of 1/10° with a resolution of 5 cm. In addition classification is further aided by employing sector–scan techniques. For close range identification an even higher resolution capability will be required.

In the future new and improved transducer materials will be required to increase cavitation limits. The development of synthetic aperture sonar processing would enable the range and coverage rate of MCM sonars to be increased without degrading resolution and requiring a large underwater body for their deployment. Parametric sonar techniques would enable sonars to penetrate the seabed to detect buried mines, a feature which is now of paramount importance in the face of the latest mine technology and the growing importance of littoral warfare. The potential threat to mine countermeasures vessels from modern sophisticated mines is such that previous generations of ROV fitted with HF sonars no longer provide the necessary safety radius for the parent vessel. The next generation of ROV must, in effect, be a self–propelled sonar – a vehicle carrying both HF and LF sonars which can swim well ahead of the parent vessel, sending back to it all the necessary data for detection, classification, and identification.

As targets become smaller and more elusive the ability to detect and classify accurately will demand even higher frequencies and smaller beamwidths, while identification will be aided by optics and laser systems.

There is also the need for increased capabilities in the field of real–time data processing and post-processing techniques. Finally the MMI interface has to be improved, together with computer-aided detection and classification, which will greatly assist the operator in his task.

5.2 Europe & Scandinavia

FRANCE

DUBM 42 (IBIS 42)

The DUBM 42 side scan sonar can detect and classify mines at ranges up to 200 m with an area coverage of 2,000 m^2/s. The system is designed for use at an operational speed of 10 knots and a height above the seabed of 30 m. The basic system consists of one vehicle mounting three sonars (two lateral and one front-head) on the sonar body, one sonar display, one tactical display and one high-density magnetic recorder. The data collected by the three sonars are recorded for analysis in a land-based processing centre. In addition a visual display of the seabed is provided on the operator's console. A magnifier provides a more accurate classification. All navigation functions such as piloting and automatic control are integrated on the same console. The two lateral and single front-head sonars operate on their own individual frequency, the swept channel being 400 m wide.

TSM 2021 (DUBM 21)

The TSM 2021B (French Navy designation DUBM 21B) is a modern hull-mounted minehunting sonar which integrates with precision navigation equipment and a mine neutralisation system such as the PAP 104. The sonar carries out: mine detection at distances up to 600 m; and mine classification at distances up to 250 m through the study of target shape, echo and acoustic shadow. The sonar provides surveillance over ±175° in sectors of 30°, 60° and 90° in search mode, and 3°, 5° or 10° in classification mode. Elevation is variable, between −5° and −40°. The sonar operates on a frequency of 100 kHz for detection with pulse duration of 0.2 or 0.5 ms and a beam of 1.5° using 20 channels. Range scales are 400, 600 or 900 m. The classification sonar uses two range scales of 200 or 300 m with 80 channels and a beamwidth of 0.17°. The main components of the system comprise an acoustic transmitter/receiver transducer, associated transmission and reception electronics, and a display console. A subassembly performs the stabilisation, steering, and retraction of the detector and classifier arrays. The electro-acoustic subassembly comprises: the classification sonar transmitter/receiver chain cabinet; a display console which controls the transmission and reception functions providing a CRT presentation of target detection and location by the detection sonar; a display console with CRT displays for presentation of target locations within the classification chain; and a bridge repeater display showing range and bearing of designated targets. The TSM 2021D is an upgraded version of the TSM 2021B comprising two 19 in high-resolution colour consoles, an electronics cabinet and two acoustic antennas. The fully digitised sonar uses CAD/CAC (Computer Aided Detection and Computer Aided Classification) processing.

TSM 2022

The TSM 2022 is a lightweight (900 kg), hull-mounted minehunting sonar designed specifically for small- and medium-sized mine countermeasures vessels. The sonar can also be used for mine avoidance on minesweepers. The system integrates with precision display and navigation equipment and a mine countermeasures system. The main feature of the TSM 2022 is its small-size retractable single array assembly, used for both detection and classification, which enables easy installation and maintenance. The TSM 2022 uses high-resolution beam-forming techniques developed from the TSM 2021 together with improvements related mainly to digital processing technology. The main functions of the sonar are: detection and classification of moored and bottom mines; detection at distances up to 600 m; and classification of targets through the analysis of the shape of their echo or shadow at distances up to 250 m. In the classification mode the horizontal beamwidth is 7°, while in the detection mode either 14° or 28° beamwidth can be selected. In both modes the vertical beamwidth is 15°. The main units comprise: the hoisting and stabilisation system together with the retractable array which is installed in the sonar trunk; the electronics cabinet; and the operator's console. The console provides display of sonar images in the various modes, memory and display of former sonar contacts, and includes interfaces with current mine disposal and navigation equipments. In the Mk II configuration the TSM 2022 is equipped with a 19 in high–resolution colour console and specific sonar processing such as computer–aided detection (CAD) and computer–aided classification (CAC) as well as a performance indicator (PI).

TSM 2023

The TSM 2023 hull–mounted minehunting sonar has been developed by Thomson Sintra in collaboration with Simrad Subsea of Norway. The sonar performs simultaneous mine detection and classification and comprises: a detection sonar which covers a bearing sector of 90° electronically steerable through 360°; a classification sonar which performs shadow analysis at three different frequencies with beam resolution and bearing sector being adjusted according to the size of the detected target; and two sonar consoles, one each for detection and classification and equipped with CAD and CAC processing and a performance indicator.

TSM 2050 (DUBM 41B – IBIS 41)

The TSM 2050 (French Navy designation DUBM 41B) is a high–resolution side scan sonar system designed for the location and classification of targets lying on the seabed. It consists of three towed sonar vehicles, two consoles, the necessary cables, shipboard hoists and handling gear. The 340 kg system is designed for use by low tonnage vessels, and in waters of 100 m or deeper. In normal operation only two vehicles are streamed, the third being kept as a ready spare. The towed vehicles operate at about 5.5 to 7.5 m above the seabed at speeds of 2 to 6 knots. The tow cables are provided with deflector vanes which ensure that the bodies are towed at a distance to port and starboard of the ship's track, and marker floats are provided to indicate the line of travel for each sonar. The streamlined vehicle is fitted with the side scan sonars, plus additional sonar transducers for determining its height above the seabed and for obstacle detection. Servo–controlled fins are provided for depth control, and roll stabilisation is achieved using a separate set of four fins. An acoustic pinger is installed as an aid to recovery should the vehicle break its tow. Manual controls are provided at the operating console to permit manual piloting, either to override automatic control or as a standby mode. Each of the two vehicles operates on its own frequency, the two frequencies being separated by 50 kHz, both sonars scanning the area of seabed between the vehicles. A permanent record of the sonar data gathered is made on a facsimile type recorder and a tape recorder. The information is also simultaneously presented on two CRT, these console–mounted displays providing an image of the seabed. A typical scanned area covered by the two bodies amounts to a total width of about 200 m with maximum sonar range selected. There are two other range settings, 50 and 25 m, and the latter setting resolution is stated to be better than 5 x 10 cm in the lateral (scan) direction and in the line of travel.

TSM 2054

The TSM 2054C is a high–resolution multibeam side scan sonar performing detection and classification of mine–like objects at a maximum speed of 15 knots. The swept channel width is 200 m. The towfish sonar vehicle is 3 m long and weighs less than 300 kg. The vehicle can be navigated manually or automatically at a constant depth between 6 m and 200 m with an altitude in relation to the seabed of from 4 m to 15 m, programmable from the sonar console keyboard. It can automatically follow seabed terrains with gradients of more than 15 per cent with high precision. The vehicle also carries an obstacle avoidance sonar that allows it to 'overfly' 10 m high obstacles at its maximum towing speed. The vehicle is also fitted with a doppler sonar for accurate speed measurement. Automatic correction of speed, yaw and pitch, and the precise location of the towed body is computed aboard the ship to an accuracy of 9 m and less than 1 m in altitude. The operating speed of the vehicle in relation to water is 4–15 knots. Two sonar ranges are provided – 2 x 50 m and 2 x 100 m. The resolution is 2 x 10 cm or 2 x 20 cm, depending on the selected sonar range. The sonar provides coverage up to 5.4 km^2/h. The system provides two operating concepts: data collected by the sonar are recorded for analysis in a land-based processing centre; alternatively detection and classification of new mine-like contacts is carried out on board ship. The system is based on a multifunction display console.

GERMANY

DSQS-11 M

The DSQS-11 M hull-mounted minehunting sonar is a high-definition, dual-frequency sonar which forms an integral part of the Atlas MWS 80 minehunting weapons system. The sonar performs: detection and classification of all types of ground, short tethered and moored mines; high-clearance performance by simultaneous detection and classification within a 90° sector; simultaneous 3D precision target localisation; computer-aided detection (CAD) and classification (CAC); automatic tracking of stationary and moving targets; integrated sonar performance prediction function; integrated route survey (side scan function); and video and data recording and playback. The system comprises an operator console, an electronics cabinet, the hoisting unit (with transducer arrays, signal processing units and stabilisation unit) and the hydraulics unit. The console incorporates two high-resolution raster scan displays and keypads for direct system operation. Additionally, the operation is supported by easily accessible menu guidance on both graphic displays.

ITALY

SQQ-14IT

The SQQ-14IT minehunting sonar is based on the American AN/SQQ-14 system but incorporating new electronics for both the dry and wet ends. Only the mechanical parts of the hoist system have been retained. In line with the next-generation sonars, new functions such as real-time performance monitor, speed scan, memory display and remote display have been included, as well as BITE functions which provide self-diagnosis and the location of malfunctions down to the printed circuit cards. The mounting of the transmitter unit in the towed body has provided an increase in the system operational performance, as well as a redundancy of the principal functions, that is, console interchangeability and the possibility of operating with reduced efficiency in the case of malfunction in one or more of the transmitter sections. The adoption of the latest consoles for search and classification operations has greatly improved the MMI, and simplified the use of the system and the training of operators. Recording and mission playback facilities are available, as well as training and simulation programmes.

NORWAY

SA950

The SA950 is an active, high-resolution, multibeam, sector scanning mine detection and avoidance sonar system. The design ensures simple and accurate operation, a high degree of maintainability, and low size and weight. The sonar comprises a hull unit (with transducer array), transceiver unit, servo unit, and an operator/display unit. Detection is achieved using 95 kHz frequency combined with high source level and CW or FM signal processing. Transmission is sectoral, covering a 45° or 60° sectors, with 32 beams each of 1.7°. The sector is mechanically trainable and tiltable covering ±190° in the horizontal plane and from +10° to −90° in the vertical plane. The centre beam is stabilised against pitch and roll. Signal processing is performed in a fast digital system using the full dynamic range of the echo information. The receiver has time variable gain and automatic gain control on the preamplifiers, background normalisation after the beam-forming, and ping-to-ping correlation. The echoes are presented in mono or 64 colour high-resolution on a 20 in monitor. There is sector presentation and time/range (echogram) presentation. In addition, PPI, relative or true motion is selectable and target or position can be tracked. The system is controlled by a single operator using designated hardware switches, joysticks and trackerball on the operator unit front panel and interactive with a menu on the display. The Mk 2 variant can be used as a stand-alone system or form part of the Micos integrated command system. It can also be used in combination with Simrad towed array sonars. The sonar interfaces to doppler log, gyro, radio position system, VRU, CTD sensor and tactical system.

UNITED KINGDOM

Type 162M

The Type 162M side-looking, solid-state sonar fitted in ships of the UK Royal Navy, detects and classifies both mid-water and seabed targets.. Operation is simplified by entirely automatic gain control. The three 49.8 kHz transducers produce a fan-shaped beam pattern about 3° in azimuth and 40° vertical; with the side-looking elements having axes 25° below the horizontal. Three range scales are available: 0 to 300; 0 to 600; and 0 to 1,200 yds, and accuracy is better than two per cent assuming a sound velocity of 4,920 ft/s. The sonar displays port and starboard recordings simultaneously on a single straight-line recorder. The paper speed changes automatically when the range scale is selected, and the speeds are 6, 3 and 1.5 in/min (152, 76 and 38 mm/min) respectively. At a ship speed of 10.8 knots the display scales are the same across the paper and vertically. A fix marker draws a line across the width of the paper when a button is pressed, and it can be operated in conjunction with a Type 778 echo-sounder or other equipment.

Type 193M

The Type 193M 860 kg, solid-state, short-range, high-definition sector scanning uses digitally processed video and display systems, together with computer-aided target classification. The sonar operates at two frequencies, providing detection and classification. The sonar provides both bearing and range data; the fine range and bearing resolution enabling the operator to assess accurately the shape of a target and hence its nature. Since the resolution depends both on operating frequency and pulse length, two selectable frequencies are provided, with a choice of pulse lengths in the classification mode. Range and bearing data appear on two displays: in the search mode one display shows the total range covered; in the classification mode a 27 m section of the search display is expanded to fill the second display screen and permit close examination of the target. The sonar uses two frequencies: 100 kHz for long-range search and 300 kHz for short-range search and classification. The transducers are carried on a stabilised, steerable mounting, the whole assembly contained within a dome. The receiver uses a modulation scanning technique with 15 beams, 1° wide in LF and 0.33° wide at HF, giving azimuth coverage of 15° and 6° respectively. The surveillance of wider areas of the sea bottom is achieved by means of an automatic search sequence selected in accordance with the type of sea bottom. The returned echoes are presented to the operators on separate CRTs at the control console. One of these displays range and bearing of targets within the sector being scanned by the search transducer, while the other is used for expanded range presentation in search mode and also the presentation of the classification channel data. The controls for adjusting the transducer position, signal parameters such as frequency and pulse length and for co-ordinating with the rest of the minehunting and destruction systems, are also provided at the console. Type 193M data can also be fed into other ships' systems. The Speedscan system can be added to the Type 193M to allow it to be operated in the side scan mode and to generate a hard copy printout of the seabed. Range and bearing data outputs are provided in synchro and digital forms, allowing the sonar to be interfaced with various plotting tables, action information systems and remote indicators. The system is also fully compatible with the Nautis-M.

Type 2093

The multimode 13,000 kg Type 2093 sonar system is designed to operate in either hull-mount or variable depth mode, deployed through the ship's centre well. A dual-frequency search and dual-frequency classification capability enables the sonar to operate under all bottom and sea conditions at a range claimed to be twice that of hull-mounted sonar and with an increase of three times detection depth. The Type 2093 is integrated with the ship's command system to allow simultaneous operation in combinations of search, detection, classification and route survey modes. The towed body of the 2093 is cylindrical with hemispherical ends. Stabilising vanes consist of fixed strips mounted a few centimetres away from the shell, largely parallel to its surface and curved so as to be concentric with its axis, and with the centres of the hemispherical ends over an arc of 45°. The sonar arrays within the body consist of two 360° rings for LF and VLF operation, hydrophones for HF and VHF reception, and separate HF and VHF projectors. Below is a cylindrical VLF projector and finally a VLF depth sounder. LF and VLF are used for search, survey and moored mine classification; HF and VHF for classification and survey. The Type 2093 employs multiple operating modes, which offer the MCM commander the choice of a number of operational configurations to match the prevailing threat and environment. These multiple operating modes comprise VLF and LF search and moored mine classification, VLF and LF search, search and ground mine classification, also search and route survey. The modes can be combined to enable concurrent operation.

Type 3010/T (Minescan)

Minescan (3010/T) is a low-cost route surveillance and mine detection side scan system particularly suited to craft of opportunity (COOP) applications, and for use by reserve or regular forces. Key elements of the system are: heavy-duty, dual-frequency towfish, 100/325 kHz, operable to 300 m depth, with optional depressor; dual cabinet transportable system in shock-mounted, splashproof units; single cabinet system for permanent installation applications; colour video display with image enhancement and target marking facilities; range selection - 50, 100, 200 and 400 m port and starboard, and with a total swathe width up to 800 m; image correction - slant range and speed over ground; high-density sonar data recorder with random access target image recall; high-resolution thermal linescan recorder; and interfacing to integrated navigation processors.

MS 58
The MS 58 hull-mounted minehunting sonar is a low-cost, lightweight dual-frequency system. It uses modern high-speed processing modules presenting a flexible system architecture which allows the system to be offered in a number of forms: mine avoidance sonar for minesweepers or non-MCM vessels; standard single-operator minehunting sonar; or standard dual-operator minehunting sonar. All variants have facilities for search and classification of ground mines, and special array and processing features for the detection and classification of moored mines and surface and near-surface mines. A full route survey capability is incorporated as a standard feature. A modern lightweight directing gear provides good platform stability characteristics with small hull aperture requirements. A number of optional facilities are available including CAD, and multiping processing and CAC.

Sea Scout
Sea Scout is a lightweight, portable, rapidly deployed shallow water minehunting/avoidance sonar, which can also be used for detection and classification of seabed ordnance in shallow water conditions to a range in excess of 300 m. It is designed to operate from small COOP offering rapid response in high-threat areas. The sonar is based on a single high-frequency (250 kHz) mirror system, and carries out a continuous search of the seabed, detecting and classifying objects in the water column or on the seabed. Contact range and bearing are relayed to any command system fitted on the parent platform to geographically position contacts for further investigation. The mirror array offers optimum in-water beam-forming with a minimal number of components. It also allows the system to be used at greater speeds than conventional MCM systems. The mirror provides a fixed azimuthal field of view of 20° with a narrow azimuthal beam that is electronically scanned over the full field of view to provide an azimuthal resolution of nominally 0.6°. Sea Scout is capable of being mechanically scanned ±30° to provide an overall azimuthal field of view of at least 80°. The system has a fixed vertical field of view of ±5° selectable within the total vertical range of +10° up to −45°, this being set as a function of the operational mode. In mine avoidance mode the operator sets the centre of his vertical field of view nominally around the sea surface to a specific depth in the water column, for the detection and classification of contact mines. In seabed search mode, the operator simply sets the depression of the sonar system to provide optimum cover of the seabed as a function of the water depth. The array can also be turned through 90° in order to provide a height-finding capability. In all operational modes, target data can be transferred automatically to the command system by the operator positioning a cursor over the detected target and pressing a button.

PMS 75 Speedscan (Sonar 2048)
The PMS 75 Speedscan is a self-contained side scan modification for ships already fitted with a minehunting sonar system. It allows the sonar system to be operated in a side scan mode and generates a hard copy printout of the seabed. When the sonar is operating with Speedscan, its transducer is trained 90° to port or starboard of the ship's track. As the ship proceeds along a predetermined track, an area of seabed parallel to the track is interrogated by the sonar. The received sonar data and reference timing signals are fed to the Speedscan processor which forms a side scan beam and presents the processed data as hard copy on continuous recording paper. The equipment comprises two main assemblies, each contained in a portable case. In use, the two cases are stacked and locked together with the end covers removed. The upper case houses the processor and the operator's control panel, the lower case the recorder unit. Speedscan presents information on light-sensitive paper exposed in the recorder. On the record, the seabed is represented as a series of parallel narrow strips of selected length, perpendicular to the ship's track. The sonar data from the whole azimuth sector scanned by

the sonar is extracted by Speedscan to represent a narrow strip of seabed along the sonar beam axis. These data are extracted by the beam–former to generate a parallel beam of sufficient width to overlap the strips covered by the two previous transmissions, so providing the three-ping sample of each point of the seabed along the range axis. The beam–forming system is programmed by ship's speed and sonar range to provide a controlled interrogation of the seabed. This information is used to intensify and modulate a light source in the recorder which produces a high–definition image of the seabed.

5.3 North America

CANADA

CMAS–36

The lightweight CMAS–36 hull–mounted mine detection and avoidance sonar has been developed with the emphasis on detecting tethered mines in shallow waters. It is a dual frequency (36 kHz/39 kHz), high–speed scanning, multibeam sonar using narrow receiving beams (6° horizontal and vertical) and digital processing techniques to provide instantaneous detection, classification and target data. Video presentation on a 16 in colour display in either PPI or B–scope is available in both active and passive modes, a trackerball–controlled cursor providing target range, depth, bearing, speed and heading for instantaneous display. Selectable omni or directive CW transmissions on 270 channels are available in the active mode with selectable transmit sector width and position. The automatic detection mode outlines one or more of approximately 100 detection zones in which signal reception exceeds the operator-selected threshold. The surveillance zone is fixed at 72° centred on the bow. The sonar has been offered for the Canadian mine warfare programme.

Model MS 992

The MS 992 side scan sonar is a simultaneous dual– or single–frequency sonar designed for route survey and minehunting aspects of MCM. A unique aspect of the MS 992 is its two-wire telemetry that allows integration of the subsea electronics package to almost any cable configuration. The side scan is manufactured in both conventional towfish and ROV versions. Towfish are built in either negative stainless steel or neutrally buoyant packages. Standard transducers are available in 120/330 kHz frequency packages for both long–range and high-resolution performance. Trials with the Canadian Navy have consistently produced 200 m range per channel at 300 kHz. Other frequency options are available. Digital sonar data (12 bit) is output through a standard SCSI interface which allows the transfer of data for storage, external analysis and to drive digital hard copy recorders. Image enhancement is achieved with the powerful floating point 32 bit DSP processor. The video display of sonar data allows complete flexibility in presentation – including a true data expansion feature that can be used in real time without affecting the full range of data being stored to tape, optical disc or hard copy. Playback of recorded data also permits use of these features. The compact surface processor is designed to operate four wire 971 scanning sonar heads as well as the MS 992 side scan electronics. This offers advantage in terms of equipment commonality and system flexibility – especially when the side scan is ROV–mounted. To ensure repeatable sonar records, a critical requirement of any geophysical survey, the MS 992 automatically calibrates the system each time it is turned on. The side scan outputs both raw sonar data and all critical system setting parameters to tape, including optional sensor package information and navigation data.

UNITED STATES

AN/SQQ-14

The AN/SQQ-14 variable depth towed mine detection system is a solid-state, dual-frequency sonar for simultaneous detecting and classifying bottom mines in shallow water. It features a towed body in the shape of an elongated sphere towed through a centre well on a minesweeper. The AN/SQQ-14 can operate down to 45 m transmitting at 80 kHz and 350 kHz for search and classification respectively, scanning over azimuths of 100° and 18°, and through an elevation of 10° in each case. Azimuth and range resolutions are 1.5° and 1 m for search, and 0.3° and 8 cm for classification. The sonar uses beam steering. Separate search, classification and memory display consoles are provided.

AN/SQQ-30

The AN/SQQ-30 variable depth towed mine detection sonar is the successor to the AN/SQQ-14 system. The system comprises two sonars: a search sonar for initial detection; and a high-frequency, high-resolution sonar for classification of targets. The two sonars are separated to give the area coverage needed. They are housed in a hydrodynamically shaped vehicle and towed at various speeds by the minehunter. The towed body is streamed from a well in the forward part of the ship, using a winch driving a 3 m diameter cable drum on the foredeck. Two display consoles are provided for the search and classification sonars.

AN/SQQ-32

The AN/SQQ-32 advanced minehunting search and classification sonar system consists of two separate sonars: a search sonar for initial detection; and a high-frequency, high-resolution sonar for classification of the targets. The two sonars are partially housed in a hydrodynamically shaped vehicle and towed at various speeds by the minehunting ship. The latest technologies in beam-forming, signal processing, modular packaging and displays are used throughout the system. The detection sonar operates over a wide range of distances and bottom conditions. This part of the system incorporates a CAD facility which is designed to help the sonar operator to discard non-mine objects detected by the sonar. The classification sonar is based on the French DUBM 21 and TSM 2022 minehunting sonars. It provides very high-resolution transmission and reception of underwater signals to enable targets to be identified with near-picture quality. The 'wet end' consists of a hydrodynamically shaped vehicle and a deployment/retrieval system. The towed body is housed in a vertical trunk extending from keel to deck, just forward of the bridge. The body is roughly egg-shaped, with a pair of boomerang-shaped vane arms pivoted on either side. When stowed in the trunk, these lie alongside the body itself. Once clear of the ship's keel they swing aft to stabilise the body as it is towed along in the direction normal to its axis. The body is slung from pivots on either side to a fork assembly that is connected to the towing cable, so that its axis remains vertical. Within it are the search sonar staves arranged in a 360° ring array belted around the body, together with immediate sonar electronics and associated equipment. Below the body is the rotating scanner for the classification sonar which can operate at three frequencies down to 300 kHz. The AN/SQQ-32 consists of two identical operator consoles with high-resolution displays. Search and classification data can be displayed simultaneously or independently. A search display presents video patterns for the detection operator to examine, and computer-aided detection indicates objects likely to be mines. Search data are presented on six screen displays allowing the operator to determine if a mine-like object is a real target or a natural bottom feature. The system proceeds to long-range classification with the second operator, measuring the detected object's height above the seabed in order to assess the probability of it being a moored mine. The classification operator then uses the high-resolution classification sonar, which has very narrow beams with dynamic focusing, to examine objects detected on the bottom. Automated detection and control functions are performed by two AN/UYK-44 computers.

SH100

The SH100 hull-mounted mine detection and classification sonar is an active, dual-frequency, high-resolution sonar. The sonar is designed to operate from the amphibious landing zone out to depths of 100 m for detecting and classifying bottom mines and beyond that for detecting moored mines. The SH100 is an extension of the single frequency SA950 mine avoidance sonar. The transducer arrays are mounted on a fully retractable hull unit which is mechanically stabilised against roll, pitch and yaw. The signal processing, including beam-forming, is performed in a fast digital signal processing system using the full dynamic range of the signals. The SH100 operates at 95 kHz (LF detector ranges to 600 m) and 335 kHz (HF classifier ranges to 200 m) and has a sectoral transmission and multibeam reception covering an LF sector of 45° with 32 beams of 1.6°, and an HF sector of 16° with 64 beams of 0.25°. Vertical coverage is 10°. The sectors, which are aligned along the same axis, are mechanically trainable (+-200°) and tiltable (+10 to -90°). Mine detection with the SH100 has been demonstrated at ranges greater than 1,000 m. The sonar can operate on both the LF detection and HF classification frequencies simultaneously. Signal enhancement and display techniques, such as shadow mode normalisation, FM pulse compression, ping-to-ping filtering, echogram detection and zoom, can be applied separately to each frequency. The following operational modes are available: wide sector search and target acquisition using only the LF; high-resolution classification and precise localisation using only the HF; combination of LF and HF on alternate transmissions; and vertical positioning of the transducer for moored mine classification and depth measurement. The SH100 comprises a hull unit, preamplifier, transceiver, servo control unit, servo transformer unit, and a control and display unit. The basic configuration includes software and hardware interfaces to doppler speed log, gyrocompass, roll and pitch sensor, and surface navigation systems. The control and display console includes two high-resolution 20 in RGB monitors mounted one above the other. The lower monitor displays the sonar echo image while the upper one displays tactical information. The echo display presents the sonar image in B-scan and/or echogram modes. The tactical display presents a map-like image of the sonar situation including the ship symbol and past track, target positions and classifications, LF and HF sector coverage, and tilt and train graphics. Additional tactical data displayed include the ship data such as position, course and speed, and target positions (both relative and geographical). The simple MMI provides access to all the primary controls with the buttons and joysticks on a control panel. Secondary commands are accessed through an on-screen menu. Enhancements to the basic configuration are available providing data distribution to multiple locations on the vessel, recording of target database, sonar image printing, and more powerful tactical and navigational functions. The sonar can also integrate with other Simrad systems such as: vessel control system; electronic chart display and information systems (ECDIS).

Model 5952

The Model 5952 is a lightweight (less than 100 kg) dual-beam, high-resolution side scan sonar suitable for deployment from COOP. It uses simultaneous dual-frequency insonification of the target for multispectral data collection and processing. The simultaneous dual-frequency insonification technique is used for target discrimination because targets reflect various frequencies in different manners, thereby providing the operator with the opportunity to observe the reflectivity of difficult targets with both high and very high frequencies. The returns of both frequencies are displayed simultaneously on either hard copy, very high-resolution thermal paper or on a high-resolution video display unit (VDU). The sonar operator has a direct, real-time visual comparison of the target returns for optimum evaluation and target selection. The sonar system uses a variable depth sonar transducer which can be selectively deployed to avoid interference from thermoclines. Combined 100/500 kHz frequency for simultaneous insonification of target areas, and 3.5 kHz for penetration of the sea bottom are available for use in the determination of bottom hardness for use in the assessment of a buried mine threat. The sonar images are transmitted up the tow-cable to a combined sonar transceiver and graphic recorder for processing and display on high-resolution thermal graph paper. Alternatively, the combined sonar image can be displayed on a high-resolution digital video display unit. Performance enhancing accessories are available which permit advanced image processing of the sonar data, as well as data reduction. Fully integrated sonar systems, which interface with navigation and shipboard equipment, are also available. The VDS transducer is easily deployed by a single crewman without specialised handling equipment.

Multiscan Mine Countermeasures Sonar System

The MultiScan side scan sonar uses digital technology to form five dynamically focused, collimated sonar beams on each side of the sonar, each with a 20 cm along-track resolution out to approximately 80 m with an ultimate beam angle of 0.15°. The design permits contiguous placement of the collimated sonar beams allowing gap-free, 100 per cent bottom coverage at a speed of advance of 10 knots. This capability is achieved while maintaining the specified 20 cm along-track resolution. As a consequence of the generation of the multiple side scan sonar beams, the sonar can produce up to five times the data of an unfocused, single beam side scan sonar at significantly higher area coverage rates. At slower speeds, the sonar beams are controlled to maintain equally spaced, contiguous beam placement, so that the sonar maintains 100 per cent bottom coverage. Multibeam coverage increases the probability of detection for difficult mine-like targets by providing increased target data over that of a single beam sonar. The MultiScan system was designed to be deployed in the towed mode, as the VDS design permits the transducer to be towed in the vicinity of the sea bottom for optimum resolution and for the avoidance of environmental interference. However, for shallow water operations the MultiScan can be hull-mounted and maintain operational levels necessary for MCM operations in water depths as shallow as 2 m. The system consists of a towed variable depth sonar transducer, a tow cable, and a surface control module. The latter is fitted with a high-resolution thermal printer for archival display of sonar data or offers as options high-resolution digital video display of sonar data as well as digital data recording on 8 mm cassettes. The sonar operates at a nominal frequency of 380 kHz for the optimum combination of operating range and target resolution. The pulse repetition rate is operator controlled for selection of optimum operating range.

System 2000

System 2000 is a lightweight (less than 65 kg) simultaneous dual-beam (100 kHz and 500 kHz) side scan sonar system which includes integrated target analysis and data storage for COOP MCM applications. The simultaneous dual-frequency insonification technique is very beneficial against modern mines that employ shapes and materials to make detection by sonar difficult. The system has both a very high-resolution (300 dpi) hard copy thermal printer and high-resolution video display. The sonar returns of both frequencies are displayed simultaneously on the hard copy printer along with relevant navigation and status information. The high-resolution video displays the real-time sonar data and through the use of the integrated trackerball, true target zooms, mensuration, and position can be made. This target information/position can be logged off to a file on the digital tape storage unit, as well as output to an external printer. Selectable and programmable colour palettes are available to enhance bottom features and targets. System 2000 uses a variable depth sonar deployed via a single coaxial tow cable. The fully digital towed sensor with integral dual–frequency transducers features a built–in multiplexer and provides 12 bit data resolution. Programmable pulse length, tone burst transmitters coupled with a very low–noise front end maximise range performance and noise immunity. This design architecture results in a system performance which is suitable for very shallow water MCM applications as well as deeper operating depths. Optional sensors are available such as heading, depth, roll, pitch, responder and so on.

Q–MIPS Processing System

Q–MIPS is a fully integrated, menu–driven sonar data collection and image processing system which is compatible with any side scan sonar. It provides digitising, processing, display, colour hard copy and permanent storage of sonar images. Various configurations are available to meet a range of MCM tasks including minehunting, Q–route surveillance, COOP operations, range and harbour security and geophysical survey. Q–MIPS is compatible with any side scan sonar analogue output and features: automatic range scale setting, trigger detection and bottom tracking; four simultaneous channels of the highest speed 12–bit A/D; very high sampling rates with the ability to change sampling schemes to user requirements; 256–colour waterfall display with real–time corrections for slant range, water column, radiometric non–linearity and speed; shipboard colour hard copy; permanent safe storage on optical disks for high capacity and instant retrieval in full resolution; automatic navigation interfaces; spatial image processing enhancements including many edge detection and filtering algorithms, plus unlimited zoom and rotation; transform domain image enhancements, including DFT filters. Sonar images are processed at over 80 Mflops.

TABLE 5.1 SONAR EQUIPMENTS

System	Manufacturer	Country	Installed on	Navy	Units
AN/UQS-1		USA	MSC 268/292	Iran	3
AN/UQS-1		USA	MSC 268	South Korea	3
AN/UQS-1		USA	MSC 289	South Korea	5
AN/UQS-1		USA	MSC 294	Greece	9
AN/UQS-1		USA	Adjutant/MSC 268	Spain	8
AN/UQS-1		USA	Adjutant/MSC 268	Taiwan	5
AN/UQS-1		USA	Adjutant/MSC 268/ MSC294	Turkey	11
AN/UQS-1		USA	Bluebird	Denmark	1
AN/UQS-1		USA	Bluebird	Thailand	2
AN/SQQ-14	Lockheed Martin	USA	MSC 322	Saudi Arabia	4
AN/SQQ-14	Lockheed Martin	USA	Adjutant	Greece	6
AN/SQQ-14	Lockheed Martin	USA	Aggressive	Spain	4
AN/SQQ-14	Lockheed Martin	USA	Aggressive	Taiwan	4
AN/SQQ-30	Lockheed Martin	USA	Avenger	USA	6
AN/SQQ-32	Raytheon	USA	Yaeyama	Japan	3
AN/SQQ-32	Raytheon	USA	CME	Spain	4
AN/SQQ-32	Raytheon	USA	Avenger	USA	8
AN/SQQ-32	Raytheon	USA	Osprey	USA	12
CMAS 36	C-Tech	Canada			
DSQS-11H	STN ATLAS Elektronik	Germany	Bang Rachan	Thailand	2
DSQS-11M	STN ATLAS Elektronik	Germany	Bay	Australia	2
DSQS-11M	STN ATLAS Elektronik	Germany	Frankenthal	Germany	12
DSQS-11M	STN ATLAS Elektronik	Germany	Hameln	Germany	10
DSQS-11M	STN ATLAS Elektronik	Germany	Lindau	Germany	14
DUBM 20B	Thomson Sintra	France	Circe	France	5
DUBM 42	Thomson Marconi Sonar	France			
MG 7		Russia	Yevgenya	Angola	2
MG 7		Russia	Yevgenya	Bulgaria	4
MG 7		Russia	Yevgenya	Cuba	12
MG 7		Russia	Yevgenya	India	6
MG 7		Russia	Yevgenya	Russia	27
MG 7		Russia	Yevgenya	Syria	5
MG 7		Russia	Yevgenya	Vietnam	2
MG 7		Russia	Yevgenya	Yemen	5
MG69/79		Russia	Natya I	Ethiopia	1
MG69/79		Russia	Natya I	Yemen	1
MG69/79		Russia	Sonya	Bulgaria	4
MG69/79		Russia	Sonya	Cuba	4
MG69/79		Russia	Sonya	Ethiopia	1
MG69/79		Russia	Sonya	Russia	62
MG69/79		Russia	Sonya	Syria	1
MG69/79		Russia	Sonya	Vietnam	4
MG69/79		Russia	Vanya	Bulgaria	4
MG69/79		Russia	Vanya	Russia	13
MG69/79		Russia	Vanya	Syria	2
MG79/89		Russia	Notec	Poland	17

5.4 Market Prospects

A brief glance at Table 5.1 shows that apart from Russia (which is a special case), France undoubtedly leads the world in the export of MCM sonars. Although only known fittings have been listed in the Table, there is little doubt that the market leader is the former Thomson Sintra Company which, since it has integrated itself with Ferranti Thomson Sonar Systems and GEC–Marconi Sonar Systems Division in the UK under the name of Thomson Marconi Sonar Ltd, must surely now be the world's largest company involved in developing and manufacturing sonars for mine warfare. Exactly what impact this will have on the market remains to be seen, for having only recently been formed, the new grouping will need time before it can rationalise its strategy regarding sonar development, marketing and so on.

On the French side the Table lists a total of 86 ship sets (5 x DUBM 20B, 35 x TSM 2021, 36 x TSM 2022, 4 x TSM 2023, 6 x TSM 2054), although actual numbers manufactured and sold exceed this, as a total of 60 TSM 2021 systems have been ordered. By far the greatest number of systems (at least 71) have been exported to 14 countries. Of the 86 systems known to be on order or installed on specifically named ships, only 18 have been supplied to ships built in France.

The picture which has emerged shows that not only has the former Thomson Sintra Company captured the world market for MCM sonars, but that it has not been limited to installing these sonars on vessels built in France. Of the systems installed in vessels built in other countries, 17 have been supplied to ships built in the Netherlands, 11 to Sweden, 8 to South Korea (as retrofit to upgrade US–supplied vessels), 7 to Belgium, 6 to Denmark, 6 to Turkey, 4 to Norway, 4 to Taiwan, 3 to the USA for ships built for Egypt, and 2 to Yugoslavia.

With this impressive export record it is unlikely that France's position as world leader will be greatly challenged in the near future – especially under the new grouping of Thomson Marconi Sonar. Apart from Russia and Japan, France now has an entree with all the world's shipbuilders involved in MCMV construction.

As well as its own personal record in the field of MCM, the former Thomson Sintra Company has also supplied sonar technology to Simrad Subsea in Norway and the Raytheon Company in the USA. In the latter case the two companies have jointly developed the AN/SQQ–32 VDS sonar of which a total of 27 systems are on order. Of these 7 systems are for export – 3 being installed in the Japanese *'Yaeyama'* class and 4 on order for the new Spanish CME.

Two other European countries apart from Russia have a major capability in MCM sonar – STN ATLAS Elektronik of Germany and the Sonar Systems Division of the British GEC–Marconi Company (now part of Thomson Marconi Sonar). With regard to GEC–Marconi, two main systems have been developed and manufactured for both the domestic and export markets. The older Type 193 system which is now reaching obsolescence and the latest generation Type 2093. Of the Type 193, a total of 25 systems are currently operational – the system having originally been installed in the UK Royal Navy *'Ton'* class minehunters which have now all been decommissioned. A total of 13 systems remain operational aboard the UK Royal Navy *'Hunt'* class MCMVs. but these will be replaced as part of the *'Hunt'* class mid-life upgrade. Of the remaining 12 systems, 6 are installed in South Korea's *'Swallow'* class, 4 in South Africa's *'Ton'* class, and 2 in Argentina's *'Ton'* class.

The latest Type 2093 is beginning to have an impact on the export market, and to date a total of 23 systems have been ordered, a half of which will be installed on the UK Royal Navy's *'Sandown'* class. Of the remaining 11 systems, 3 are for the Saudi Arabian *'Sandown'* class, 6 for the Australian *'Huon'* class currently under construction, and 2 for the new MSC 07 boats building in Japan for the JMSDF. This latter sale is of some significance for it is the first time that Japan has acquired a sonar system that has not been supplied by a US company or developed domestically from a US system. If this proves a success Britain may well have broken the US hold on the Japanese sonar export market. Japan devotes a not inconsiderable percentage of its naval budget to MCM and this potentially large market could present significant opportunities for British sonar systems in the future.

In Germany STN ATLAS Elektronik is the prime supplier of sonars for MCM. To date the bulk of the DSQS 11 sonar production – 36 systems – have been sold for domestic use. Four systems have been sold for export, 2 to Australia and 2 to Thailand.

Of the other countries involved in manufacturing sonars for MCM, Italy builds an upgraded version of the US SQQ–14 sonar primarily for domestic use (12 systems on the *'Lerici'* and *'Gaeta'* classes) and Norway the SA 950 for the 5 *'Alta'* class. Norway has also sold 4 x SA 950 systems to Turkey – but it is not known on what class the system has been installed. Systems have also been sold to other navies but no details are available.

The only other European country to have built large numbers of sonars for MCM is Russia. These have all been built for domestic construction under the old Soviet Union, with large numbers being exported to satellites and other friendly countries.

Of the systems built – most of which must now be considered obsolete and ineffectual in the face of the modern mine threat – five main types remain in service on seven classes of ship serving with 15 countries. While Russia has certainly built and exported many more MCM sonars than any other country, these are nearly all mounted on obsolete ships. Furthermore, most of the countries operating these ships are unlikely to be in a financial position to either replace them or upgrade them with new sonars. The two main exceptions to this generalisation are India and China. Both countries are developing their own indigenous MCMV construction and sonar capability and should be in a position to begin meeting their own requirements by about the year 2010.

As for Russia herself, new MCM sonars are under development but no details are available, and in view of the parlous financial state of the country, development is probably very delayed. Construction of new MCMVs and installation of new MCM sonars and retrofit programmes lag very much behind any planned schedule. Any possibility of Russia re–entering the export market in MCM sonars in the near future seems very unlikely.

While the overall market for new build systems in the immediate future is relatively limited, there are a number of medium to long term projects planned which will require new sonar systems. Among the countries seeking to acquire new MCMVs are Brazil, Bulgaria, China, Greece, India, Indonesia, Japan, Kuwait, Portugal South Africa, South Korea, Taiwan, Thailand and Turkey.

Of more immediate interest is the retrofit/upgrade market for MCM sonars. Here, Belgium, Denmark, France, India, Italy, Malaysia, the Netherlands, Singapore, South Korea, Sweden, the UK, and USA will all need to address this problem by the turn of the century.

SECTION 6 – UNDERWATER MINE WARFARE VEHICLES

6.1 Introduction

The task of mine countermeasures is becoming ever more difficult as mines become more sophisticated. It is now recognised that MCMVs must remain well outside the danger zone of any potential mine. Dealing with the modern influence mine is an extremely complicated affair requiring a very careful approach. Once an object has been detected and classified as a mine by a minehunting system, it has to be neutralised. In the past this highly dangerous task was accomplished either by the MCMV itself – an operation of considerable danger to the ship concerned – or by clearance divers operating from inflatables carried by the mine countermeasures vessel (MCMV).

Ever since the Second World War, shallow water minehunting has been carried out using hull–mounted minehunting sonars mounted on the MCMV. These sonars were designed to scan ahead of the vessels searching for, detecting and classifying objects at what was then considered to be a safe distance ahead of the ship. However, because of the position of the sonar in relation to the seabed (the angle of incidence of insonification) and the problems of acoustic propagation near to the surface where thermal layers can create barriers to the sound propagation, these sonars were really only effective in depths down to about 100 m.

To overcome some of the problems associated with hull–mounted sonars, and in particular for operation in deeper waters where mines began to be laid, the variable depth sonar (VDS) was developed. This allowed the sonar to penetrate the inhibiting thermal layers, and to be placed much closer to the level of the potential target. However, the VDS is limited by the fact that being a towed body it necessarily trails in the water, a characteristic which increases with the vessel's speed and the depth of immersion of the body. The towed body could therefore finish up quite some distance behind the towing platform, placing the latter in a vulnerable position in relation to the danger zone of any mine. To overcome this limitation necessitated the sensor having improved detection and classification ranges. These characteristics could only be achieved by use of a lower frequency, which in turn resulted in loss of resolution and contrast, features essential in minehunting.

Using technology originally developed for the offshore industry, a range of remotely controlled underwater vehicles has been developed to undertake mine neutralisation tasks. The advent of the remote controlled mine disposal system (RCMDS) has given mine countermeasures and vessels used on MCM tasks a much greater flexibility of operation. Using the RCMDS, MCM operations can now be carried out under much more adverse conditions than would have been the case if only mine clearance divers were available. Moreover, the operation itself is completed more quickly and there is a greater certainty of success without the risk of loss of life, or damage to the ships.

As the safety and efficiency of minehunting platforms is now regarded as a high priority, so the sensors used to detect mines must realise greatly improved performance characteristics. One of the technologies now being examined by a number of navies to overcome the limitations of both the hull–mounted and variable depth sonars and the limitations of current RCMDSs in the face of modern mines, involves deploying a remotely operated vehicle fitted with a highly sensitive minehunting sonar some considerable distance ahead of the parent MCMV platform. This is frequently referred to as the 'dog–on–lead' principle.

To carry out its role effectively, the underwater vehicle must be capable of fulfilling a number of functions, the most important of which are: it must be able to approach an object classified by the MCMV sonar as a possible mine target without the possibility of triggering the target and so endangering itself and the MCMV; carry on board sensors and relay data back to the MCMV for further evaluation of the target; on positive identification carry out a neutralisation attack under remote control (to lay a mine destruction charge near the mine).

In addition to these requirement the vehicle must be capable of operating in a wide variety of environments ranging from shallow to deep water, sandy to rocky bottoms, clear water to very muddy water, and low or high sea states in all areas of the world.

To develop an underwater vehicle to carry out the tasks outlined above in such a hostile environment requires the very highest degree of technological development covering a wide variety of disciplines such as mechanical and electrical engineering, metallurgy, computer technology, hydrodynamics, and so on. It is not surprising, therefore, to find that firstly there are very few companies in the world involved in development underwater vehicles for mine warfare and secondly that they and the equipments carried by the vehicles are at the forefront of technology.

An idea of the magnitude of the problems facing designers of underwater vehicles can be gauged from some of the following features which must be attainable in any such designs. Namely: they must be easy to handle, deploy and recover from the water, even in adverse weather conditions; the operator in the MCMV must be able to guide the vehicle to the target in the shortest time possible; the vehicle must be capable of allowing the operator in the MCMV to carry out positive identification of the target; to carry out its role effectively the vehicle must not generate any form of signature (magnetic, acoustic or pressure) which might trigger a mine; the vehicle must be extremely manoeuvrable, particularly at virtually zero speed in all planes, in order that the operator can carry out detailed inspection of the target.

To meet these criteria underwater vehicles all exhibit similar characteristics but differ in specific details. Thus, all are constructed of non–magnetic materials with a hydrodynamic shape and are capable of carrying some form of countermining charge. Propulsion systems do differ by are usually either electric motor or hydraulic drive. Some form of sensor must be carried. In the past this has often been only a TV camera, but in the presence of limited visibility and for positive identification it is now recognised that to be fully effective the vehicle should also mount a small high resolution sonar.

The vehicle must be connected to the MCMV by some form of umbilical link through which the operator can control the movements of the vehicle, and via which the data gathered can be sent back to the MCMV. This requires that on the MCMV the operator's console should possess a TV monitor, ppi presentation of the vehicle's movements, some means with which to control the movement of the vehicle (usually a joystick), and presentation of the vehicle's sonar data. Alphanumeric tote display for data from the vehicle is also advantageous.

Generally, operation of underwater vehicle entails the operator guiding it down the beam of the MCMV's sonar to the vicinity of the target. When a satisfactory range is reached the operator switches on the vehicle's own TV and sonar and guides it onto to the target. Once in close proximity to the target the operator manoeuvres the vehicle around the target at very slow speed and close range. Having carried out a detailed examination of the target the operator manoeuvres the vehicle to a suitable position near the target and sends a coded signal to release the countermining charge. The vehicle is then recovered and the charge detonated by coded signal or time delay fuze.

Usually, underwater vehicles can only carry out one mission, as most are unable to carry more than one countermining charge. Furthermore, they usually only have sufficient onboard power for one mission.

Mine development, continues to advance, and in the future new technologies will be incorporated into future generations of underwater vehicles making them far more autonomous, enabling the parent vessel to stand off at much greater distances from danger areas, allowing the vehicle to carry out search, location and identification, tasks previously carried out by the MCMV.

6.2 Europe & Scandinavia

FRANCE

PAP Mark 5

ECA's fifth generation of the PAP 104 family, the 104 Mark 5, is able to simultaneously hunt and destroy ground mines and tethered mines by either placement and remote detonation of a charge or by cutting the mooring rope of tethered mines. The very low–acoustic and magnetic signature Mark 5 incorporates new features as well as those retained from previous versions. In addition, two new options have been added to the PAP; an umbilical cable and CLAD, a new device to destroy tethered mines. The autonomous PAP carries its own source of energy – a sealed lead–acid battery which can be recharged or changed on board between missions. The vehicle is remotely controlled via a fibre optic cable from a control console situated in the operations room. The wire is consumable and dispensed by the vehicle. The PAP is horizontally propelled by means of two side thrusters. Vertical thrusters are used to provide mid–water navigation. Using a guide rope, the vehicle can navigate at constant altitude, the length of the guide rope being varied to allow fine adjustments of vehicle altitude during a seabed mission. The front section of the vehicle can optionally mount either a high resolution near field sonar or a mid–range relocation sonar of the customer's choice. Identification is carried out by a tiltable low–light TV camera and/or by a tiltable colour TV camera. The PAP Mark 5 carries a 126 kg charge for mine destruction and cutters can be carried with the charge. A manipulator with associated camera and projector can be carried instead of the charge. The vehicle also incorporates a radio control for surface recovery in case of cable breakage.

GERMANY

Pinguin B3

The STN ATLAS Elektronik unmanned, remote controlled Pinguin B3 is fitted with a 1,000 m fibre optic cable for data transmission. Using data derived from minehunting sonars, the vehicle is able to run almost automatically towards the target. In low current conditions the operating range is up to 900 m at 200 m depth, while in currents of 3 knots the range is about 500 m at 200 m depth. The Pinguin provides different sonar options and is also fitted with a tiltable light–sensitive CCD underwater camera. The vehicle identifies the target and, if a mine, releases a disposal charge to destroy the target. A second charge enables the vehicle to continue the mission and approach a second target. The vehicle can also be fitted with an anti–moored mine device. The propulsion system consists of two variable speed horizontal propeller thrusters, laterally mounted on the stern section of the vehicle. At low speeds depth control is achieved by an automatically controlled vertical thruster mounted at the hydrodynamic neutral point of the vehicle, at higher speeds (>2 knots) diving is dynamic using flaps. Power is provided by an internal battery system. Control is exercised by the operator feeding inputs on the control console for heading, depth and thrust. The vehicle can be manoeuvred freely within the whole water column and ranges of the minehunting scenario.

Seafox & Seawolf
These two one–way mine disposal vehicles are being developed by STN ATLAS Elektronik under the framework of the German Navy's MA 2000 MCM improvement programme. The concept is based on the use of remotely operated unmanned surface drones (Seahorses) towing different sonars and one–shot vehicles (the Seawolf for mine destruction). Specific requirements which the Seawolf has to meet are that it should be capable of destroying all types of mines, including the deployment of a large blast charge to destroy buried mines using shock waves, and a relocation capability in time–dependant destruction operations. This will form part of the later upgrade of the *'Frankenthal'* class minehunters. Also being developed under a shorter time scale, as part of the upgrading of the German Navy's MCM forces, is a new role for the minesweepers of the *'Hameln'* class. Within this part of the programme five boats will be converted into advanced minesweepers and five into minehunters similar to the *'Frankenthal'* boats, deploying the Seafox. Seafox is a smaller, lighter and less expensive version of the Seawolf, but will incorporate many of the components and subsystems of Seawolf. Full–scale development of Seafox began in 1993. Seafox is a one–way mine disposal vehicle capable of ensuring the destruction of moored mines at varying depths, and short tethered mines, using a shaped charge. The vehicle is basically a self–propelled wire–guided ammunition with a homing capability, controlled from the ship platform and normally operating within the range of minehunting sonars. The vehicle will be capable of carrying out mine destruction in a very short time (less than 10 minutes) using automatic guidance with a manual override capability. Seafox comprises five main subsystems: the one–way vehicle; control unit; storage and launch facilities; data transfer cable; and interfaces to the MCM platform. Both upgrade programmes are relatively short term, and are due to commence in 1997, with Seafox scheduled to enter service in 1998.

INTERNATIONAL

PVDS
The Propelled Variable Depth Sonar (PVDS) system developed by Thomson Sintra ASM of France in co–operation with Bofors Underwater Systems (SUTEC Division) of Sweden, is a third–generation minehunting sonar concept which seeks to meet the increased requirement for safety and efficiency in MCM. The aim of the PVDS concept is to provide a full minehunting capability covering survey, detection, classification and identification of all types of mine threat throughout the underwater environment, from a propelled underwater vehicle. It is intended that this will assure the complete safety of the MCMV by providing maximum range of detection and classification in front of the parent platform, without compromising the probability of detection and classification of the various types of threat. Being a modular concept, the PVDS has the flexibility to offer adaptation to all existing and new types of MCMV platform. Trials off the Swedish coast in September 1995 demonstrated that this concept is particularly valid in areas where varying thermal layers create a mirror effect (in this instance at a depth of some 12 m, about 6 to 8 m above the seabed). In such conditions the 'dog–on–lead' concept allowed the sonar to be navigated in the layer where the threat was anticipated, and well ahead of the parent MCMV, ensuring it remained well outside the danger

area of any mine. The trials showed the PVDS was capable of being accurately navigated near the seabed, maintaining a satisfactory performance from the surveillance sonar. In these conditions the PVDS provided a detection range on Manta and Rockan mines (two of the most difficult types to detect) of about 300 m at a distance of some 500 m ahead of the ship. In other words an overall range between ship and mine of some 800 m. However, due to the proximity of the PVDS to the seabed, shadow classification was not as clear as in normal conditions due to the very shallow grazing angle and multipath propagation. The production version of PVDS will feature improved performances including the incorporation of a third LF for long–range detection over a 90° sector; simultaneous detection and classification; computer–aided detection and classification; and a x2 increase in resolution for VHF classification and HF detection. The PVDS concept is based on a self–propelled vehicle linked to the MCMV by an umbilical cable. This cable provides power to the vehicle and also incorporates a fibre optic cable which is used to pass the full range of sonar data, which its sonar gathers, to the MCMV for further processing. To assure full safety and efficiency, the PVDS can navigate in combination with its parent platform at a speed of 5 knots, at depths between 5 and 300 m and at a distance of 200 m in front of the MCMV, closing to within about 10 m of the target for classification. These characteristics have enabled the designers to mount a shorter–range high–frequency sonar in the vehicle offering improved detection and classification performances. Furthermore, the ability to manoeuvre in three dimensions enables the vehicle to place the sonar array in the optimum position, irrespective of the speed of the MCMV. Such features can offer a considerable increase in mission effectiveness, placing the vehicle close to the threat to provide a highly accurate classification. In 1990 Thomson Sintra began development of a sonar to meet the above requirements. The aim was to develop high integrated front–end electronics using hybrid technology which could be located inside the array beam without the need for an extra electronic container. In addition dual–frequency transducers were developed to perform detection and classification resulting in a slim acoustic array just 0.7 m in length exhibiting low drag. These developments led to a sonar weighing just 80 kg, small enough to be fitted to a medium–size ROV. The vehicle selected to carry the sonar was the Double Eagle, a fully proven and highly manoeuvrable, low–magnetic and acoustic signature underwater vehicle. The resulting PVDS design has realised a platform with much better performances than a VDS, at 500 m in front of the MCMV and using a detection frequency of 165 kHz (as opposed to the 30 to 80 kHz of a VDS) a bearing resolution of 3.6 m (compared to the 11 to 30 m of the VDS) is obtained. The sonar is the TSM 2022 Mk 3 with yaw and pitch mechanical stabilisation which is mounted at the front end of the Double Eagle. Detection performance of the 165 kHz frequency sonar offers a search sector of 63° (mechanically steered) and a range on a sandy bottom of 500 m, an angular resolution of 0.7°, range resolution of 12.5 cm, and vertical beamwidth of 19°. The swept path is 300 to 400 m. The classification performance of the 405 kHz frequency sonar provides a 12° search beam which is electronically steered within the 63° classification beam operating to a range of 160 m, an angular resolution of 0.3°, range resolution of 6.25 cm and vertical beamwidth of 7°. For moored mine classification the vehicle is rolled through 90°.

SVDS (MWS 90)

STN ATLAS Elektronik of Germany in co–operation with ECA of France is developing the SVDS based on the PAP 104 Mk 5 RCMDS and the DDQS–11M sonar, demonstrations of which were scheduled to take place in 1996. The system is designed to operate at a speed of 8 knots and to provide sonar data down to depths of 300 m, deploying ahead of the parent vessel to counter ground, short tethered, long tethered and self–propelled mines. The sonar is based on the STN ATLAS DSQS–11M family and comprises a high–resolution, dual–frequency system. In addition the SVDS vehicle houses some of the signal processing electronics. The 1.4 m linear array for horizontal detection and classification is mounted in a horizontally rotatable sonar wing mounted beneath the body of the underwater vehicle at the rear on a horizontal axis for 90° horizontal detection/classification. The depth array is mounted behind a sonar dome in the head of the vehicle and allows for simultaneous depth classification of contacts within the horizontal search sector of the linear array. To ensure that the antenna orientation is aligned with the planned track, and to overcome current influences on the vehicle, the linear array can be operated in a stabilised mode. The depth array normally follows automatically the orientation of the sonar wing and can additionally be freely trained within the detection sector. The sonar is controlled from a single console with sonar data acquired by the parent vessel's hull-mounted sonar and the SVDS sonar displayed on the two displays of the console. The display includes the 90° horizontal detection sector with two windows for classification and depth evaluation. The operator's task is aided by CAD/CAC techniques and target data correlation between the hull-mounted and the SVDS sonar data. Software performance evaluation tools to assess and optimise parent vessel missions and/or SVDS operation are integrated and can be used by the operator either for mission planning or during missions. By combining sonar and vehicle performance prediction under actual measured environmental conditions, mission effectiveness can be optimised. The integrated sensors and acoustic positioning system enable various vehicle control modes to be achieved, including constant height over the seabed, constant depth, automatic track keeping and manoeuvring and automatic winch control and tether management. To provide for virtually unlimited mission duration the vehicle is powered via an umbilical cable from the parent vessel. The cable includes fibre optic cores for the transmission of sonar data and vehicle control data.

ITALY

MIN Mk 2

The Whitehead Alenia MIN (Mine Identification and Neutralisation) system, has the capability to identify and neutralise both bottom and moored mines. The latest version, the Mk 2, comprises: a self-contained, hydraulically powered, wire-guided submarine vehicle; a main console in the operations room for vehicle guidance and control; a tracking system for autonomous localisation of the vehicle; a portable auxiliary console for guidance and visual control of the vehicle during launch and recovery operations; and a set of operational accessories including - an auxiliary battery charging station, and a device for oleopneumatic power pack recharging. The Mk 2 vehicle is powered by a closed-circuit oleopneumatic accumulator, enhancing the low-noise profile and non-magnetic characteristics. A steerable main propeller allows the vehicle to approach the target detected by the minehunter's search and classification sonar. In addition, the employment of horizontal or vertical thrusters gives the vehicle a high manoeuvrability and hovering capability when submerged and high manoeuvrability when surfacing for the recovery phase. Vertical manoeuvring, even to the maximum depth of the vehicle, is controlled through water ballast tanks which are filled or emptied by means of pressurised air; hydraulic power is consequently used only during transfer from/to parent ship and identification of the target. TV camera and sonar are orientable through 150° in the vertical plane allowing flexibility of observation during both bottom and moored minehunting. The system can use two types of weapons (bottom charge and explosive cutter) thus enabling the hunting of either bottom or moored mines. The main console is of modular configuration and consists of separate processing units and operator desk to minimise possible installation problems. The desk on the main console is fitted with a control lever and joystick for speed and direction control of the vehicle during the search and approach phases, and for fine adjustments of the vehicle's position when it is in close proximity to the target.

The main console also incorporates the controls for the actuator of the high–resolution sonar in the MIN and for its tracking subsystem. Two 9 in video display units are mounted on the operator's desk, a black and white TV display and a high–definition display. The high–resolution sonar allows both identification of targets in conditions of poor visibility and a reduction in mission time by means of its target relocation capability. The mechanical sectoral scan sonar uses a specially developed transducer array using a diced–array technology. It comprises two major sub–units: the control and display unit which is integrated in the main console, and the underwater unit which is mounted on the vehicle and which consists of a sonar head and electronic assembly. The sonar image is displayed in either PPI presentation or B (linear) presentation. A profiling inform function is also available. The MIN Mk 2 and its control console is integrated into the MCMV combat system via serial datalinks with the command centre, parent vessel's search/classification sonar and parent vessel's gyrocompass. In addition the MIN Mk 2 is fitted with an acoustic tracking system comprising an ultra–short baseline transducer mounted on the MCMV and a transponder on the vehicle. The system processes the data relating to the target under investigation, data from the combat information centre and the classification sonar, ship's data, vehicle data, vehicle position (from tracking and/or from the search sonar) to enable the operator to steer the vehicle towards the target following the optimum track and to manoeuvre the vehicle during transit from and to the MCMV.

Pluto

The Gaymarine Pluto is available in three configurations: battery–powered with a 6 mm diameter, 500 m long umbilical cable; battery–powered with a 3 mm diameter, 2,000 m long fibre optic umbilical cable; or remote–powered of unlimited endurance with 8 mm diameter, 500 m long umbilical cable. The vehicle is powered by five thrusters: two horizontal for forward/reverse, two vertical for vertical and lateral shift, and a transversal thruster, controlled by two joysticks. The vehicle can maintain automatic depth control within ±10 cm. Forward maximum speed is 4 knots and vertical speed 1 knot. Sensors are mounted in the forward tiltable section of the vehicle. Optional equipment includes black and white LLTV, colour TV or still camera, search or scanning sonar, acoustic pinger, strobe flash, measuring instruments, manipulators and so on. There are 10 free channels for remote control and two four–digit telemetry channels for measurements. The console incorporates a 9 in TV monitor for display of information showing TV image, depth, compass, head tilt angle, elapsed time and sonar diagram. All displayed data can be video recorded.

Pluto Plus

The basic configuration of Pluto Plus is the same as the standard Pluto version, but incorporating a number of improvements. Pluto Plus uses the latest fibre optic floating, reusable 2 km long cable and hydrodynamic design to provide improved observation capabilities, extreme stand–off range, an increase in speed to 7 knots, an increase in endurance from 6 to 10 hours (battery capacity has been doubled) and the lowest possible drag factor. The vehicle is fitted with special sonar sensors for navigation, search, obstacle avoidance and identification. The sensors are all mounted in a single package featuring ±100° tilt and ±80° pan. A single control console monitor gathers and displays the video picture, navigational data, sonar graphics and maps of the investigated area. The lightweight, compact, Pluto Plus low–magnetic and acoustic signature vehicle is resistant to shock and vibration to MIL–SPEC standards and is designed to operate from all kinds of vessels without special or expensive handling equipment.

SWEDEN

Double Eagle

Double Eagle is a comparatively low–weight system is driven by eight thrusters (two forward, two lateral, four vertical) giving over 5 knots forward, 3 knots reverse, 3 knots lateral and 1 knot vertical movement, in currents up to at least 3 knots. Double Eagle is fitted with a computerised stabilisation control system and is extremely manoeuvrable, exhibiting unlimited movement in six degrees of freedom (pitch ±180°, roll ±180°). The display comprises two monitors with control keyboard, one monitor displaying the picture from a colour TV camera, the other the picture from a monochrome TV camera mounted at the tip of a telescopic arm. Digital data such as heading, depth, pitch and roll angle, cable twist, leakage warnings, real–time clock with date and time and diagnostics are superimposed on the TV monitor. High–quality video is available using fibre optic link in the umbilical cable. The standard umbilical is very close to neutral buoyancy and is 600 m long. As an option a 1,000 m cable can be attached. The vehicle is fitted with one colour CCD camera as standard, and tilt arrangement with colour and SIT camera is available as an optional extra. Optional sensors include echo–sounder, and electronic scanning or conventional sonar. The vehicle uses a unique precision charge placement technique using a small mine disposal charge which also features several advantages. The vehicle can also carry a heavy standard NATO mine disposal charge as well as cable cutters.

6.3 North America

CANADA

Dolphin

The air–breathing, diesel–powered, deep ocean semi–submersible Dolphin, developed by International Submarine Engineering of Canada, is a UHF radio link remote–controlled vehicle which is similar to a small snorkelling submarine. Power is provided by a Sabre 212 turbocharged, intercooled marine diesel engine rated at 150 shp. The vehicle can operate for up to 24 hours at a speed of 12 knots. Dolphin has been launched and recovered without any in–water manned assistance in Sea State 5. It can be pre–programmed to follow a set course with waypoints, running autonomously. When running autonomously, the control radio can monitor mission progress. The vehicle's configuration allows for easy integration of towed and hull–mounted sensor equipment, and it features low–noise signature and sea–keeping qualities. Current sizes are in the range of 8 m in length with 150 to 300 hp installed power providing speeds of 15 to 18 knots. In water, both drone and probe refuelling is possible to provide for extended duration operations. The vehicle is capable of fulfilling a variety of missions requiring stability in high sea states, long endurance and high speed. Two missions which have been demonstrated are minehunting and hydrography. To demonstrate the hydrographic mission, a multibeam hydrographic echo–sounder was installed in 1990. With accurate positioning from differential GPS, surveys can be conducted at substantially lower costs than those requiring the use of manned launches. The minehunting operation involves real–time transmission of water column and sea bottom mine–like contacts to a controlling platform, and provides battle groups and other sea–going formations with an organic minehunting capability. To demonstrate this mission, a cable winch which could handle side scan sonar towfish was developed in 1990–1991. Two vehicles were delivered to the US Naval Coastal Systems Station Panama City in 1988, and have been outfitted with a variety of mine–hunting sensors for evaluation. One of these vehicles was converted to a Remote Minehunting Operational Prototype (RMOP) in 1994. The RMOP vehicle was outfitted with a fleet MCM towed sonar, a cable winch complete with towfish handling/docking system, a high-bandwidth RF link for real-time transmission of sensor data, a precision GPS receiver, and hull-mounted forward-looking sonar. The command and control features were integrated with US Navy consoles, and installed into a standard military control van. The RMOP integration was a joint government/industry collaboration undertaken by NCSS, Rockwell AESD and International Submarine Engineering (ISE). It has been extensively tested and operated from the *'Spruance'* class destroyer *John Young* in MCM support of a US amphibious exercise in March 1995. Two other vehicles have been used by the Esquimalt Defence Research Detachment (formerly Defence Research Establishment Pacific - DREP), Canada, for the development of a Canadian Remote Minehunting System (RMHS).

Canadian Remote Minehunting System

Macdonald Dettwiler of Canada is developing the next-generation remote minehunting system (RMHS) which will have applications to all minehunting sonar sensors, as well as to shore-based facilities. One of the applications for the RMHS as part of Canada's new MCDV programme is the concept of route survey. The concept involves the use of a sonar imagery map database of areas of the seabed which are susceptible to a mine threat, showing the position of all known objects. Using this database as a reference, the MCDV can carry out a rapid survey of a selected area, comparing a new sonar image against one previously recorded, and from which any new object can be quickly detected and where necessary investigated. This will result in a significant reduction in the time required to monitor a route and neutralise any new mine-like objects which are detected. In this system the MCDV will be able accurately to position objects in a continuous sonar map database, geocoding and mosaicing the data. The use of this geocoded sonar imagery will enable the MCM tasks of route survey, minehunting and neutralisation to be carried out independently, yet at the same time integrated by a single geographic database. The system elements comprise the MCDV, a shipboard mine warfare control system, the route survey sonar payload, the vehicle and a mine warfare data centre. For route surveying the MCDV will deploy a towed side scan sonar which will provide the necessary data to build up and maintain a sonar image of the seabed. During minehunting, a remotely controlled vehicle will tow the sonar and radio the sonar imagery to the ship's command and control centre. Mine neutralisation may be carried out by an underwater vehicle or expendable mine destructor. Each MCDV will have a permanently installed mine warfare control system comprising tactical MCM sensor positioning, data analysis and database management subsystems (DBMS). The tactical subsystem will integrate all navigation sensor inputs and generate a local operations plot, based on electronic charts. These will display 'own ship' position, sonar towfish position, radar contacts and mine warfare contacts. The data analysis subsystem includes multifunction detection and classification consoles for the display and analysis of real-time and historic sonar imagery. The DBMS retrieves all sonar imagery and contacts from previous missions for display by the data analysis subsystem. As new objects are detected and classified they are stored by the DBMS.

The primary requirement for the sonar is that it should be capable of detecting buried ground mines, mines with anechoic coatings and irregularly shaped mines in conditions of high background clutter. It will operate at depths down to 200 m, close to the seabed to provide good object shadowing for accurate classification and avoiding thermal and salinity layers. Resolution in the classification mode is 12.5 x 12.5 cm and high-coverage rates dictate a speed of advance of 10 knots with a 400 m swath width. To meet these requirements a multibeam side scan sonar has been selected. The array is housed in a hydrodynamically stable towfish body, incorporating active control surfaces for bottom following, and comprehensive position monitoring instrumentation. In the route survey mode the sonar towfish is stored in a standard 6 m ISO container module incorporating its own deployment crane and winch. During both launch and recovery the crane maintains positive control of the towfish, release and recovery taking place below the surface. The module can be installed on the MCDV within 12 hours. The route survey inspection payload comprises a towfish complete with launch and recovery system. The towfish is electro-hydraulically powered through an umbilical which gives it unlimited endurance. Manoeuvrability in three axes is provided by four thrusters. Sensors include a side scan sonar, colour and LLTV cameras. The towfish is monitored and controlled from the ship's command centre while deployed, its maximum operating depth being 300 m at a range of 600 m. The vehicle designed to tow the towfish sonar for minehunting and radio sonar imagery to the ship is the Dolphin. To carry out mine neutralisation a number of options is available including clearance divers, ROVs and expendable mine destructors. However, underwater vehicles are expensive, and the risk to divers from today's sophisticated mine is considerable. In view of these considerations, much effort is being devoted to developing effective mine destructors. These are relatively inexpensive self-propelled weapons which can be guided to their targets by sonar, detonating in close proximity to the mine with both high hard and soft kill effectiveness. The sonar payload is designed to operate down to depths of 200 m, close to the seabed. The multibeam side scan sonar is carried by the hydrodynamically stable towfish body which incorporates active control surfaces for bottom following and comprehensive position monitoring instrumentation. The geocoding and mosaicing technology applied to the sonar imagery uses algorithms developed by Macdonald Dettwiler which automatically fuse geocoded imagery from sensors of different resolution into a single image. Geocoding is sensor independent and the technology can be used to enhance the performance of sector scan sonars and to integrate side scan sonar data with that of sector scan sonars. Geocoding allows sonar imagery to be registered precisely to a chosen map projection and absolute geographic locations assigned to targets identified in the imagery. The sonar towfish ground speed, its position relative to the towing platform and the seabed, its roll, pitch and yaw, together with differential GPS and gyro navigational data are integrated into the geocoding process to position each pixel of sonar imagery in a geographic reference grid. Geocoding permits sonar images acquired during earlier missions to be compared with those of the present mission and changes detected semi-automatically. Work in progress indicates that geocoded sector scan images may also be mosaiced, either for the purpose of building up a continuous map of the seabed or to smooth out background noise to improve the detection and classification range of the sonar. Mosaicing is the process of combining swaths of geocoded imagery to create a continuous sonar image map database. Mosaicing fills in gaps in the sonar map with new mission data and discards duplicate imagery. The best data can thus be used to create the mosaic. Large areas which are in shadow from one aspect can be replaced by usable imagery acquired from a different aspect. The single mosaiced database is faster to access and the preparation of mission databases simplified.

UNITED STATES

AN/SLQ-48 MNS

The Alliant Techsystems mine neutralisation system (MNS) is designed to detect, locate, classify and neutralise moored and bottom mines, using high-resolution sonar, low-light level TV, cable cutters and mine destruction charges. The tethered underwater vehicle carries these sensors and countermeasures and is controlled from the parent vessel. Initial target detection and vehicle guidance information is provided by the ship's sonar. Initial vehicle navigation is plotted and monitored within the MNS acoustic tracking system. Vehicle sonar is used during the mid-course search and final homing phases, and high resolution enables operations in poor visibility by sonar guidance alone. Low–light TV is used in conjunction with sonar during the precision guidance phase near the target. Power to the two 15 horsepower hydraulic motors which drive horizontal, vertical and lateral thrusters is provided via a neutrally buoyant 1,070 m umbilical cable, which also carries signal and control links between the vessel and the MNS vehicle. Aboard the parent vessel sonar screens on the control console display sensor and vehicle status information in alphanumeric form. The monitor and control consoles are the focal point for operation and management of the complete system. Display facilities include: vehicle sonar, vehicle TV, deck TV, vehicle control and navigation, and provision for monitoring system status Alliant Techsystems is developing MNS II with a smaller footprint and performance improvements. MNS II integrates the separate monitor and control consoles into a single console and reduces the size of the cable handling and power system. Intended for smaller MCMVs, MNS II also provides fibre–optic telemetry and increased thrust that will raise vehicle speed to >8 knots. Using alternative sensors MNS II can extend the current MNS mission beyond bottom and moored mine neutralisation to the following applications: buried mine location and neutralisation, mine search and classification, bottom mine survey, search capability operating ahead of the parent platform and recovery or repair of underwater objects.

TABLE 6.1 ROV SPECIFICATIONS

Vehicle	Country	Length	Diameter	Height	Weight	Payload	Depth	Speed	No of Units	In Service with
		(m)	(m)	(m)	(kg)	(kg)	(m)	(kts)		
AN/SLQ-48	USA	3.8	0.9	1.2	1,247			6	67	2
PAP 104	France	3	1.2	1.3	850	1 x 126	300	6	350	14
Dolphin	Canada	7.30	0.99		3,265			12	10	2
Double Eagle	Sweden	1.9	1.3	0.8	400	80	500	5		3
Min Mk 2	Italy	3.55	1.05	1.5	1,150	120	350	6	8	1
Pinguin B3	Germany	3.5	1.5	1.2	1,350	2 x 125	200	6		2
Pluto	Italy	1.68	0.6	0.63	160	15	400	6	70	10
Pluto Plus	Italy	1.95	0.51	0.55	200	15	400	6	25	3
Seafox	Germany	1.1					300			

6.4 Market Prospects

In the face of the modern mine threat there is an undoubted need for modern minehunters, and possibly even minesweepers, to carry some form of underwater vehicle. In view of the opinion expressed earlier, route survey must now form the basis upon which all mine warfare operations are conducted. Route survey operations have to be conducted with such precision and in such detail that it will become impossible for them to be carried out without the aid of an underwater vehicle fitted with a precision high frequency, high resolution sonar.

The modern minefield, skilfully and artistically laid, with mixes of various types of mine, will present mine countermeasures forces with considerable problems. Again, the efficient and safe conduct of mine countermeasures operations will be virtually impossible without the aid of some form of underwater vehicle.

Although there are numerous underwater vehicle designs available, many from the commercial offshore market, few of these are now considered suitable for specific operations against a mine threat – although many are suitable for a wide range of other tasks such as the inspection of fixed seabed installations (signature ranges, underwater sensor ranges such as SOSUS, and so on) and the recovery of weapons from trial ranges and so on.

Underwater vehicles for mine warfare neutralisation tasks and for route survey will need to be more carefully designed and engineered to operate in the face of a modern mine threat. Even minehunting vehicles are now being targeted by mine sensors. To meet these demanding requirements only a few European and North American/Canadian companies have ventured into this highly sophisticated area of technology.

Not surprisingly in view of its domination of the MCM market, France has secured the major part of the market share in underwater vehicles dedicated to mine warfare. With total of more than 350 vehicles now operational in 14 navies, ECA has secured the claim of world leader in this field. However, this lead will be challenged in the coming years as the concept of underwater vehicle operations changes and the need for vehicles to carry more sophisticated sonar sensors comes more into focus. Hence, the joining of forces between major minehunting sonar manufacturers and specialist companies devoted to underwater vehicle development. In order to better place themselves to retain their lead, ECA have joined forces with STN ATLAS Elektronik in Germany to develop a self-propelled minehunting sonar system based on the PAP 104 vehicle and the DDQS-11M sonar. This system is being developed to face the direct challenge from the giant French Thomson Sintra company (now Thomson Marconi Sonar Ltd) with its TSM 2022 sonar currently installed in the Bofors Underwater Systems Double Eagle. The latter system is further down the development path than the French/German system, and with its current export drive with the Double Eagle the Bofors system is proving a major challenge to the ECA PAP 104 vehicle.

The other major export success in underwater mine warfare vehicles is the Italian Pluto system, with sales of 95 systems to 10/13. Pluto, however, is a much smaller vehicle with much smaller payload than the other underwater vehicles, and hence fulfils a different operating concept. Because of their size and construction they are ideal for deployment from small vessels and COOP, and are suitable for navies seeking to gain experience in MCM and minehunting operations. They also form a valuable back up to the larger underwater vehicles. In this respect Pluto and Pluto Plus are not, therefore, likely to provide a direct challenge to the PAP and Double Eagle. In their own particular field of mine warfare operations, the Pluto systems do not appear to face a direct challenge to their market prospects.

A new concept currently the subject of much discussion, is that of the expendable vehicle designed for mine destruction operations. Although Germany is currently developing its own small Seafox vehicle for this role, it would seem that if this new concept of MCM operations (which also requires a sophisticated dedicated minehunting vehicle) is accepted, then the Pluto systems will be very well placed with their long record of successful operations behind them, to capture this major new market.

The German Pinguin vehicle has primarily seen operations with the German Navy, although a number of units are operational with the Taiwanese Navy. It does not seem, however, that the Pinguin will achieve further major export sales, particularly as a new generation of vehicles is under development in Germany to supersede it in the near future.

A large number of the American Alliant Techsystem's AN/SQL–48 vehicles have been built, but the majority are all in service with the US Navy, with just a few having been bought by Japan for its *'Yaeyama'* class. Again it will be difficult for this system to achieve major export sales in the face of the very strong competition from the European companies.

TABLE 6.2 ROV INVENTORY

Vehicle	Manufacturer	Optional Equipment	Status/In Service
AN/SLQ–48	Alliant TechSystems	Sonar, TV camera,	Japan, US
PAP 104	ECA	Sonar, TV camera,	Australia, Belgium, France,
Dolphin	ISE	Sonar	Canadian Hydrographic
Double Eagle	Bofors AB	Sonar, TV camera,	Australia, Denmark,
Min Mk2	Whitehead Alenia	Sonar, TV camera,	Italy
Pinguin	STN ATLAS	Sonar, camera, MDC	Germany, Taiwan
Pluto/Pluto Plus	Gaymarine	Sonar, TV camera,	Egypt, Finland, Italy,

Note: MDC – mine disposal charge

SECTION 7 - SHIPBORNE MINESWEEPING SYSTEMS

7.1 Introduction

In many ways mechanical or wire minesweeping equipment has changed little since the Second World War, and is regarded by many as being unable to deal with the modern moored mine with its sophisticated counter sweep measures. Part of the problem lies with the fact that moored mines can now be laid in much deeper water, out to the edge of the continental shelf. Secondly, many sweep systems cannot sweep close to the seabed and cannot be manoeuvred to avoid seabed obstacles of significant height, such as wrecks, outcrops of rock and so on. There is no provision in older systems to adjust the depth setting of the sweep once it has been streamed and is in use. Finally, many older sweep systems do not incorporate a facility to monitor the sweep offset at increased speed.

There are three main methods of wire sweep operation – the single ship Oropesa sweep, the double Oropesa sweep and the team sweep using two or more ships. The former is used when the depth of the mine is known, the latter when it is unknown. Team sweeping is the faster of the two methods, but presents a greater risk to the ships. The double Oropesa sweep consists of a 200 m streamed astern of the minesweeper. At the end of this is a depressor or kite whose weight can be varied to keep the whole system at the required depth. Just above the kite stream two other cables spread out in an inverted Y fashion at the end of which are fitted a spreader or otter board (just as in the fisherman's trawl). The outer positions of these two cables and the otter which form the sweep proper are marked by floats. Attached at suitable intervals to the two spread cables are a number of cutters which operate either mechanically or use an explosive bolt. Sweeping usually involves a lead ship streaming a double Oropesa with other minesweepers following in echelon astern streaming a single Oropesa and taking up station abeam and inboard of the float streamed by the ship ahead. This ensures that no mines along the intended swept path are missed and also affords each of the minesweepers some measure of immunity, excepting of course the lead ship streaming the double Oropesa. The swept path using a double Oropesa is of the order of 275 m and the maximum depth which can be swept is about 30 m. If the double Oropesa is streamed as a team sweep between two ships, ocean sweeping depths of 90 m can be achieved with a swept path of 350 m.

The problems of wire sweeping began to be addressed in the mid 1980s when the requirements for new systems were laid down by NATO. These briefly were: an operating depth of the sweep from 10 m to 600 m or, if possible, from less than 10 m to 1,000 m, with the capability to adjust the sweep depth while remaining streamed; a system to control the minimum height of the sweep above the seabed to cut short tethered mines (minimum height 3m, ±1 m at any depth); the ability to control the depth of the sweep, or to maintain a constant depth throughout a sweeping operation; the ability to follow a bottom contour with a gradient of up to 7%; a maximum sweep speed of 14 kts and a swept path of 350 m; full operational capability (stream and recover) in up to sea state 5, and sweep in up to sea state 6; the ability to determine the precise position of the sweep relative to the minesweeper, the swept path and the depth and height above the seabed of the sweep; and finally to standardise equipment to enable it to be operated either by a single ship or as part of a team sweep.

When these requirements were set down, existing wire sweeps were fairly limited in their capability. Sweep depths at the time were limited by the tension developed in the tow wires. As the operating depth of the sweep increased, so the tension in the tow wire increased to about 130 tons at a speed of 14 kts. As the sweep descended to operate just above the seabed this tension increased to as much as 200 tons. It was also difficult to control the sweep when operating near the surface.

With regards to controlling the distance from the seabed and maintaining depth control, reliance was placed on surface floats. In team sweep mode a height of 15 m above the seabed could be maintained, but it was impossible to maintain a given height above the seabed when operating a single ship sweep. The other major problem in trying to maintain a given height was the reaction time – and this demanded a built-in autonomous level control. Similarly, problems were experienced in trying to follow a bottom contour which could only be achieved with a limited measure of success when team sweeping. The required heaving/veering speed of the winch increased as the tow speed and depth increased.

As for operational capability, systems then in use (mid 1980s) were only capable of sweeping at a maximum speed of 12 kts, with a normal operating speed of 8–10 kts – a speed considered to be much too slow to clear a route in advance of merchant shipping with current speeds of at least 12–14 kts. Again the main problem was the tension developed in the towing wire which was transferred to the winch. This in turn demanded either a larger platform with heavier winch installation to take the strain, or else a lower operating speed to reduce the tension.

These factors in turn affected the tow arrangement and hence the swept path, which was between 200 and 350 m in normal sweep conditions or limited to 50 to 100 m when sweeping near the seabed.

Since the mid–1980s various systems have been developed to meet the requirements outlined above. Developments continue, and future systems will be capable of achieving much more accurate control over sweep attitude, height and depth, and swept path, and will be able to operate at much higher speeds.

The situation with regard to influence sweeping is much more positive. Numerous developments have been undertaken and systems now available have reached a high degree of sophistication in order to counter the modern influence mine, both moored and ground laid. Developments with influence sweep systems will also continue, and particular effort will be paid to dealing with influence mines laid in the littoral zone as a barrier to potential amphibious operations.

7.2 Europe & Scandinavia

FINLAND

FIMS

The FIMS multi–influence sweep and combat control system manufactured by Elesco Oy is an integrated package consisting of all the components required for influence minesweeping operations on small– to medium–sized vessels. All elements required for a complete operation from initial planning to effectiveness assessment are provided in the package. The primary components of the system are: three 46 mm diameter buoyant electrode magnetic sweep cables, with current capacities of 400 A DC, 1,000 A repetitive peak operating at sweep speeds of up to 10 knots; various options of controllable and fixed output acoustic sources; power generation units for both the magnetic and acoustic sweep systems; a powerful shipboard control computer system, with interfaces to the ship's positioning sensors. The shipboard processing systems are located on a 486 PC computer, optimised for marine use, linked to a monitor and control out–

station which utilises a processor programmable logic controller to interface the sweep systems to the PC. For single ship operation the 486 computer is linked by cable to the logic controller system, while for remote sweeping and multi–ship operations the units have a radio link. The shipboard sweep control and positioning unit integrates control of the magnetic and acoustic sweep controllers with the ship's positioning system. This enables control of sweep line positions, displays of actual coverage, and position activated generation of sweep footprints. Both the magnetic and acoustic sweep controllers can generate complex time varying signatures. These signatures are defined interactively, and need not be a precise mathematical or transcendental function. By interfacing to the ship's echo–sounder the variation of bottom cover with depth is determined. Also the estimated sweep signal strength beneath the ship is calculated. The control algorithms optimise coverage, while maintaining a safe operating environment for the ship. Full navigation functions are provided, with the option of integral digital charting. Graphic displays are provided to show both along–track and cross–track coverage, and all position and coverage data is logged for later analysis. Historical quality control information is provided for both positioning and sweep parameters. These can be used to provide both on–line and post mission evaluations of sweep effectiveness. The window integrated navigation and sweep control and interface software includes a comprehensive training and simulation mode. The sweep coverage subsystem is one of the key features of WINS. This consists of a database of sweep coverage information organised in small geographical boxes or bins. For each bin information is stored on various sweep parameters, such as type of sweep and number of over–runs. The information is designed to be viewed graphically, which quickly reveals the success of a sweep operation, or areas where further work is required. The database is updated on the completion of each sweep line. The shipboard equipment is small enough to be readily deployed on COOP, and in this operating mode the integration of the positioning function is particularly useful, as is the integral GPS option. Its ability to use several positioning sensors eases the problems of overseas deployment or non–availability of the primary sensor. System options and variants include a containerised version, multi–ship variant, and integrated route management package, which adds a minehunting capability with the addition of a side scan sonar system.

Sonac AMS

The Sonac AMS manufactured by Finnyards Ltd Electronics, is a high–efficiency acoustic influence sweep specially designed for operation from small– and medium–sized vessels in shallow water. The system comprises the Sonac AMS deck unit, wet end unit, buoy and a tow cable. The deck unit consists of a control computer, signal generating unit, power amplifiers and matching transformers. The signal generating unit can create an unlimited number of acoustic signatures simulating any known surface vessel. Changing the signature is simply accomplished using the specially designed Sonac AMS user interface. The AMS system can be linked to external systems, for example multi–influence sweeps. The 450 kg wet end unit, which can operate down to depths of 10 m, consists of a low–frequency sound source and several optional ceramic transducers to cover the audible frequency band completely. The buoy is designed to maintain the wet end unit at a constant operating depth and can be steered via a remote radio control unit. The operating depth is controlled by a radio–controlled winch. The buoyant 500 m tow cable is specially designed to achieve high breaking strength, and the wet end unit end of the cable has an armoured section 30 m in length.

FRANCE

Sterne 1

The Sterne 1 on order for the new coastal minehunters for the Belgian Navy, is a towed influence minesweeping system which simulates both the acoustic and/or magnetic and electromagnetic signatures of various types of ship. The system comprises an AP 5 acoustic sweep and six magnetic bodies. Onboard equipment includes the TSM 2061 tactical display, the TSM 5722 sonar doppler log, and navigation systems such as the Trident III radio location system and Decca 1226 navigation radar. The Thomson Marconi Sonar Ltd system is towed at speeds of 6 to 10 knots at a minimum distance of 100 m. A mine–avoidance sonar, such as the Petrel, can be integrated into the system.

AP 5

The AP 5 acoustic sweep currently in production for the French Navy and available through DCN International, comprises: a signal generator which can generate up to 192 acoustic signatures and spectra for various types of ships, signals being pre–programmed within a wide frequency band (in both spectrum and modulation); a very high–power, low–frequency amplifier with associated impedance matching circuit; an underwater vehicle which is towed by a combined power feed and towing cable and which is fitted with an underwater electrodynamic loudspeaker with two symmetrical diaphragms; and an optional piezoelectric transducer for high frequencies. The software programs allow the generator to simulate acoustic signatures of various types of ship, to activate mines with known characteristics from a stand–off position, to activate unknown mines by scanning over their probable acoustic spectrum, to inhibit mine detonating devices within an extended area, and to transmit acoustic spectrum in synchronisation with the transmission of the current pulse of a magnetic sweep. The magnetic signature of the shipborne equipment is very low, even when operating. A pre–programmed device enables the transmitted acoustic power to be adjusted according to acoustic propagation conditions over the swept path. Minesweeping operations are performed at speeds between 3 and 12 knots (8 knots nominal) with a constant immersion of the transmitting vehicle between 8 and 10 m, which can be extended as an option. All frequency bands are covered by the sweep and electronically controlled by the computer.

GERMANY

DM 19

The 4.4 kg DM 19 explosive minesweeping cutter manufactured by Rheinmetall Industrie GmbH and in service with the German, Indonesian and Thai navies uses a linear shaped charge which assures great operational safety, even at high sweep speeds, and offers a high degree of handling safety. The lightweight equipment operates down to around 200 m depth. The cutting capacity of the system is: stud link chains 20 mm; steel wires 26 mm; and synthetic rope 40 mm. It is a one–shot system and can be mounted onto sweep wires of different diameters. No onboard maintenance or preparation for use are required. No residual parts remain on the sweep wire after detonation. The sweep wire is sandwich–shield protected.

GHA

The GHA acoustic minesweeping unit, manufactured by IBAK uses the flow of water through a turbine in the body of the vehicle to generate sound for sweeping acoustically sensitive sea mines. The GHA is towed – preferably together with a magnetic sweep – behind a minesweeper and supplies its two electrodynamic sound generators independently with power by means of water flowing over its turbines. The beat frequency, sound pressure, operating and pause time, and wobble time of the sound generator are programmable. The beat frequency can be wobbled during towing (that is the beat frequency fluctuates up and down periodically within the adjustment range). Once programmed, the operating mode is independent of the flow speed of the water, and is thus also independent of variable towing speeds. The emission of sound via the electrodynamic system ensures largely wear-free reliability of operation for several thousands of operating hours. As the system does not require a power cable the buoy is easily handled aboard ship, and is ideal for use on COOP as well as on purpose-designed minesweepers.

MSG 3

This is a single ship wire sweep system developed by the former East German Navy and used to counter moored mines laid in medium to deep water. The system can be either set to sweep at a constant depth or at a constant height above the seabed. The sweep can be deployed either as a single or double Oropesa sweep fitted with mechanical or explosive cutters. A variable depressor is fitted at the end of the tow cable and from this is streamed the sweep wire. The depressor maintains the sweep wire at a the required depth, while diverters at the end of the sweep wire maintain the sweep in a horizontal position. The diverters and depressor are almost identical and can be interchanged. The position of the diverters is marked by a surface buoy. The depth and attitude of the sweep is adjusted by means of motors which control the attitude of the stabiliser fins on the depressor and diverter, power for this being provided by a turbine-powered hydrodynamic alternator. The alternator also powers instruments which are used to control the depth, height above the seabed and angle of fins to steer the depressor and diverter. The winch on the minesweeper is computer-controlled so that it can automatically adjust the length of the tow cable according to the sweep speed, depth, tow tension and length of sweep wire. The system can automatically maintain a constant height of either 5 or 7.5 m above the seabed independent of the pre-set depth of the sweep. To avoid damage in the event of meeting obstacles of great height or gradients beyond its capability, the sweep incorporates an automatic surfacing capability. Bottom following capability is improved if a slower sweep speed or smaller swept path is employed. For example, with a sweep speed of 7 kts and a swath width of 95 m at a constant height above the seabed of 7.5 m the maximum gradient the sweep can accommodate is 8%. At maximum sweep speed of 12 kts, swept path of 220 m and with a constant height above the seabed of 5 m, the maximum gradient which the sweep can accommodate is 1%.

MSG 4

The MSG 4 single ship sweep system developed by the former East German Navy is designed for operation in shallow waters to counter short tethered mines using mechanical and explosive cutters. The sweep operates at a height above the seabed of about 2 m to sweep a path 22 m in width, being kept in this position by weights and buoys. The heavy chain sinkers, however, make the sweep complicated and awkward to handle.

Troika

The remote-control Troika minesweep system is based on three unmanned drones which are radar-tracked and remotely controlled from a ship-based operations centre. The self-propelled drones, designated HFG-F1, are equipped with minesweeping facilities which are accommodated in hollow steel cylinders encased by a ship-like hull. The propulsion system consists of a diesel-driven hydraulic power transmission and a combined rudder propeller. For optimum operation, even in rough seas, an autopilot is provided. The magnetic sweeping field is generated by two coils, mounted forward and aft on the cylinder. For acoustic minesweeping two medium-frequency acoustic hammers and one towed low-frequency acoustic displacer are provided. The remote control system uses a horizon-stabilised X/C-band precision navigation radar, a digital target extractor, a master console and three control displays. Each of the latter is associated with one unmanned drone on the one hand, and with a remote action system operator console on the other. The control displays allow the radar image of the swept channel to be greatly magnified so that a high degree of precision is achieved in guiding the unmanned drones along the specified tracks electronically inserted into the radar image. Control commands are sent to the drones in the form of multiplexed data messages, using a UHF link, which provides high reliability. Their responses and functions are monitored automatically by an integrated monitoring system. The last main feature is the reference buoy, which provides a geographically stabilised radar image and constitutes a vital component in this Troika configuration. The system is operational with the German Navy and the Netherlands Navy is studying a modified variant.

SWEDEN

SAM

Developed by Karlskronavarvet, the remote-control SAM minesweeping drone in service with the Swedish Navy uses a diesel-powered rudder propeller to drive the 18.2 m GRP catamaran hull. The craft has a range of 330 nautical miles at a speed of 7 kts. The craft is fully equipped for power supply, sweeping of acoustic and magnetic mines, navigation and remote control from a MCMV. A number of unmanned drones can be controlled, providing remote control of the diesel engine, steering, minesweeping equipment, control and monitoring. The hull of the SAM is fitted with a number of coils which are used for sweeping magnetic mines, and a towed acoustic transmitter is used to counter acoustic mines. A number of buoys is carried to mark the swept passage. A digital GPS and microplotter is also installed to monitor the sweeping track. Onboard power is provided by a Volvo Penta TAMD 70D diesel engine with a continuous output of about 159 kW at 2,200 rpm, to give the craft a speed of 8 knots. Power is fed to a Schottel-type propeller unit via reversing gear couplings and a shaft. The diesel and reversing gear are mounted on a girder framework which is shock-mounted in an aluminium structure onto the main platform. Although the machinery is normally remotely controlled from the MCMV, it can also be controlled at the engine or from the operating platform. Remote-control functions include diesel engine control, steering control, minesweeping equipment control and monitoring.

IMAIS

The Bofors SA Marine Integrated Magnetic and Acoustic Influence Sweep (IMAIS) incorporates recent technological advances in the fields of magnetic and acoustic physics as well as signal processing. The integrated sweep comprises the magnetic section, a buoyant cable with three aluminium electrodes, an acoustic generator towed at a constant depth and connected to the buoyant cable at the position of the second electrode, a control unit with keyboard and display, winch system, power pack for the sweeps – either rectifier or DC-generator, and an optional container for the containerised version. The buoyant cable's aluminium conductors supply both the electrode and the acoustic generator. The cable also has a cord with a tension load up to 10 kN, allowing sweeping speeds up to approximately 10 knots. The three-electrode system creates a unique configuration of the magnetic field. It gives a magnetically safe zone of 200 m forward of the first electrode at every depth below the towing ship. With a maximum current amplitude of 700 A (500 A RMS) the swept path is 400 m at a depth of 100 m for mines set at 100 nT. The system can be supplied for different current strengths from 75 kW to 900 kW. The magnetic field can be controlled and automatically kept in a constant and predicted configuration, independent of variations of the water conductivity. The shock resistant acoustic sweep has a maximum swept depth of 200 m. Its acoustic generator has a length of 2 m, a diameter of 0.7 m and weighs about 600 kg. It can be programmed for broadband noise and tones from a few Hz and upwards. The intensity of the sound signal can be varied, but not the frequency. The generator incorporates a depressor to ensure it is kept at the correct depth. The extremely low drag resistance of the towed sweep requires very little propulsion power from the towing ship, thereby generating very little noise disturbance and so increasing the safety of the towing ship.

The total weight of the integrated influence sweep, with streamed as well as ship–mounted equipment, is approximately 2,800 kg which makes it suitable for minehunters and small minesweepers as well as converted fishing vessels or other COOP. The system can also be supplied in a containerised version, if required. The sweep is very easy to handle and has a high manoeuvring capability in narrow waters because of its small turning radius. Compared to a closed loop system IMAIS exhibits no sag when the towing vessel slows down. The magnetic and acoustic sweep can be delivered separately.

Explosive Cutter T Mk 9
Operational with seven navies this Bofors explosive minesweeping cutter is designed to be integrated with any mechanical sweep system. To avoid activating magnetic mines, the cutter is constructed of non–magnetic stainless steel and plastic. It is so shaped to automatically take up and maintain its ready position, regardless of whether the sweep is carried on the starboard or port side of the ship. The cutter is a reloadable one shot type. On firing, the explosive device with jaws is blown away, but the fin is retained on the sweep wire and can be fitted with a new explosive device. The cutter weighs about 7 kg and is fitted with 150 g of TNT+RDX or TNT explosive. It can cut conventional chain with a diameter of 12–14 mm, stainless steel rod of 19 mm diameter, steel rod of 20 mm diameter and mild steel rod of 29 mm diameter. Sweep speed is 4 to 12 knots.

UNITED KINGDOM

Wire Sweep Mk 105
The Mk 105 low–magnetic wire sweep configuration is available in a range of sizes (including US size 4) to enable a range of naval vessels to undertake minesweeping operations. The smaller variants of the sweeps may be fitted to small patrol craft or minehunting vessels to provide a dual–role mine countermeasures capability. The larger variants are generally fitted to dedicated minesweepers or larger COOP where the space and power available is sufficient for the larger equipment. The design uses a simplified wire configuration enabling relatively small winches to be used to save deck space and in smaller sizes to reduce towing loads. The Mk 105 system may be deployed as either team, double Oropesa or single Oropesa sweeps and includes all items of minesweeping equipment necessary to fit out a vessel for minesweeping including the wires, winch, floats, cutters, kites and marker buoys. The complete systems are available in either low–magnetic or ferromagnetic materials. The system is used in conjunction with the BAeSEMA WSME system to achieve good bottom–following at depths down to 300 m.

Wire Sweep Mk 8
The Mk 8 wire sweep system in service on the UK Royal Navy *'Hunt'* class MCMVs is designed to have a low–magnetic signature in order to be compatible with the overall low–magnetic characteristics of modern mine countermeasure vessels. The Mk 8 is designed to sweep most types of buoyant mines that are moored to the seabed with sinkers by severing the mine mooring chain or wires with explosive cutters fitted to the sweep wires. The sweep system can be deployed in any of the following configurations: single Oropesa, double Oropesa and team sweep.

Wire Sweep Mk 9
The BAeSEMA Wire Sweep Mk 9, operational aboard the British and Bangladesh *'River'* class minesweepers and with the Canadian Government Naval Reserves Mine Countermeasures Project, has been developed to counter buoyant anti–submarine mines with much shorter mooring wires

laid in deep water. The system is operated in what is termed as the deep team sweep mode in which the system can be worked much closer to the seabed and at far greater depths than normal sweep systems. To operate such a system effectively the minesweeper must be able to control both the height of the sweep wire above the seabed and its flatness, so that it does not drag along the seabed and break or prematurely explode the wire cutters connected to it. The Mk 9 was adapted from the existing UK Mk 3 wire sweep system, but using a single wire with the sweep wire attached directly to the end of the kite wire. In this arrangement the kite only has to depress one wire instead of two, which enables the sweep to go much deeper with greater control. When used in conjunction with the Wire Sweep Monitoring Equipment (WSME) and the ship's echo-sounder, it is possible to heave or veer the kite wire to the exact amount required for the sweep wire to follow the contour of the seabed. Although the primary operational mode is team sweeping, it can also be deployed as a double or single Oropesa sweep. It is designed to be deployed from either purpose-designed minesweeping vessels or ships taken up from trade – usually stern trawlers. Although usually operated with two ships, several ships have been linked together to form a multi-ship team sweep providing greater swept paths and allowing an odd number of vessels to be used. Each vessel carries three lengths of wire on a winch. The outer barrels carry a sweep wire connected to a kite wire and the centre barrel carries a kite wire for Oropesa sweeping.

Wire Sweep Monitoring Equipment (WSME)

The BAeSEMA WSME was developed primarily for use as part of the Wire Sweep Mk 9 system to improve its accuracy of sweeping and is operational with that equipment. WSME enables the wire sweep to achieve a flat sweep profile at a steady seabed clearance in order to prevent the wire sagging and possibly touching the seabed and damaging the sweep, or hogging and possibly missing some mines. Previous methods of minesweeping used vessel speed to indicate sweep flatness but the errors due to ships' logs, tidal flows and current profiles often lead to sweeps with significant hog or sag. Sea trials with sweeps established that most of the tension measured as the sweep wire passed through the fairlead on the ship to the water could be attributed to the drag of the sweep in the water and the forces acting on the kite. From the measurement of this tension the speed of the sweep can be calculated, and hence its flatness. This led to the development of the (WSME) which displays digital readings of average and peak load on the sweep monitor console. The ship speed is then either manually or automatically adjusted to obtain the correct tow tension. WSME also electronically measures the length of kite wire passed through the tension meter to determine the depth of the sweep wire and displays a depth reading which can be directly compared with the digital output of the ship's echo-sounder. The winch is then either manually or automatically heaved or veered to maintain the required seabed clearance. WSME is also available with sweep location equipment for monitoring actual ground covered by the sweep. Links to the ship's command and control system are also provided.

MSSA Mk 1

MSSA Mk 1, operational aboard the UK Royal Navy *'Hunt'* class MCMV, is an acoustic minesweeping system which can generate a wide range of target-like acoustic signatures to activate all acoustic mines, including those with frequency selective triggering characteristics which are designed to be actuated only by certain types or classes of vessels. Its acoustic output is continually monitored and controlled to ensure that the required amplitude at seabed level is maintained, regardless of variations in acoustic propagation conditions. The system comprises a Towed Acoustic Generator (TAG), a Towed Acoustic Monitor (TAM) with a towed hydrophone array, and an onboard control console. The system has been specially designed to withstand the repeated levels of explosive shock likely to be experienced in operation, and is normally deployed in association with a magnetic sweep to provide a very effective method of sweeping combined influence mines.

Combined Influence Sweep CIS
By combining the MSSA Mk 1 towed acoustic generator with an efficient magnetic sweep, BAeSEMA offers a Combined Influence Sweep (CIS) system which is capable of sweeping the most modern microprocessor-controlled mines. In the CIS both the acoustic and magnetic signatures are controlled by a microprocessor-based device, giving complete and total control of the generated signature. Both target setting mode and mine setting mode techniques may be used with CIS.

7.3 Asia, Pacific & Australasia

AUSTRALIA

AMASS
The Australian Minesweeping and Surveillance System (AMASS) developed by Australian Defence Industries is a systems approach to minesweeping, providing a portable magnetic and acoustic range, combined influence and mechanical sweeps, minesweeping control system and mission planning software. The sweeps are self-contained and able to be effectively deployed from both MCMV and auxiliary minesweepers. With AMASS now operational the Australian Navy has been able to introduce the COOP concept into its force structure for use as auxiliary minesweepers.

The influence sweep is a modular, distributed dipolar system based on combinations of positively buoyant permanent magnets known as Dyads and water-driven acoustic generators, which are used to emulate the influence signatures of degaussed vessels. The acoustic generator produces a ship-like acoustic signature in the low, audible and HF ranges. The Dyads are available in two sizes and can emulate both degaussed and undegaussed vessels of various sizes up to 100,000 dwt. The Dyads are extremely rugged and manoeuvrable, making them ideal for use in confined waters and inland waterways as well as offshore operating areas. The portability of the sweeps, their ease of deployment and lack of any requirement for power or control from a towing vessel makes them an extremely flexible and cost effective system.

The portable range encompasses DC and AC magnetic, acoustic and pressure influences and can be deployed within an area of operations in a matter of hours. It provides the capability for check-ranging MCMVs as well as the ranging of fleet units and commercial vessels used to deploy AMASS sweeps in the MCM role. For non-ferrous hulled vessels the range is complemented by a permanent magnet degaussing system.

The Dyad sweeps can be supported by sweep design, effectiveness and mission planning software operating on a laptop computer. The software provides for automatic calculation of sweep configurations, to emulate ship signatures obtained through magnetic ranging, and manual design. If range data is not available the software enables both manual design of sweeps or selection of an appropriate configuration from a sweep library. Sweep signatures can be modelled against selected mine logic and sensitivity settings to establish actuation probability profiles for the selected sweep configuration. A mission planning module is also available which provides for tactical planning based on the unique characteristics of the Dyad emulation sweeps.

The lightweight and portable mechanical sweeps can be configured as team, double oropesa or bottom following sweeps. A specially designed twist–proof wire is used to reduce hog and sag in the sweep wires. The team sweep is a traditional configuration with a number of innovative features that increase the efficiency and effectiveness of the sweep. A gravity kite is used rather than a depressor, providing simple deployment and a tight turning circle. A drogue is used and levels of lift and drag are calculated to minimise sag and ensure an optimum sweep wire angle. The sweep is fitted with explosive cutters. The double oropesa sweep adapts modern fishing technology with lightweight otters and a kite that can be easily handled over the side without special davits. The bottom following sweep is a variation of the oropesa sweep, but instead of explosive cutters a catcher is used to drag mines clear of any channel. A sled, weights and floats are used to ensure the wire maintains a constant height over the seabed. The sweeps are self–contained and can be effectively deployed from COOP.

The portable minesweeping control system integrates with differential GPS and the Syledis RPS. The system can be fitted to COOP used to deploy the AMASS sweeps, providing them with the precise navigation and control capability normally associated with a purpose–built MCMV.

The precursor drone is a computer–controlled vessel providing an acoustic or combined influence precursor sweep. Monitoring of operations and updating of mission profiles can be conducted from the shore or from selected vessels at ranges up to 20 nautical miles.

The route surveillance system uses a US Klein Associates 595 towed side scan sonar and a video display processor with optical and magnetic disc recording facilities for post–mission replay and analysis. The system is interfaced with the minesweeping control system for survey operations and precise navigation. A route survey database system ashore consists of image processing and geographic information systems for exploitation of recorded data. The image processing system allows replay of sonar data from a number of recordings, and the results from several surveys of the same area can combine to provide a single, more accurate map of the area. The geographic information system provides enhanced data management capabilities including digital chart preparation, mission planning and post–mission analysis. It interacts with the image processing system to form an integrated planning and analysis facility.

CHINA

Type 312
The Type 312 marketed by the China Shipbuilding Trading Company is a small, remotely controlled mine countermeasures drone system for use in harbours and estuaries. It uses a 20.94 m long craft with a displacement of 39 tonnes which is controlled by radio signals from a shore station or mother ship up to 3 nautical miles away. During transit the drone is powered by a 12V150C 500 hp supercharged diesel engine, giving a maximum speed of 11.6 knots. During minesweeping operations the drone is powered by an electric motor. It has a range of 108 nautical miles at 9 knots. Acoustic and magnetic sensors are used but it is not clear how located mines are destroyed.

7.4 North America

UNITED STATES

EMD
The AlliedSignal Inc EMD, currently in development, is a small (length: 105 cm, diameter: 20 cm), self–propelled, expendable mine destructor for the destruction of bottom, moored and

floating sea mines. Once a target has been detected and identified using minehunting sonar, the 28 kg EMD is programmed by the operator to attack and destroy the target. The EMD is tracked by the sonar operator, and, if required, controlled via an acoustic link. Operator control commands override the EMD's internal programming for maximum control and effectiveness. The device is fitted with four shaped charge warheads which are symmetrically angled and designed to detonate the mine's explosive charge. In the event of this not being achieved, the detonation of the EMD is sufficient to disable the mine's triggering mechanism. EMD can be handled easily and launched from even small platforms. It is designed to be used with a mine-hunting sonar for kill verification, but can be used independently with expendable bottom transponders. A recoverable reconnaissance/training version with a TV camera and optical fibre link is available. The device achieves a speed of 15 kts and has an operating depth of 200 to 300m. Endurance is 15 minutes at 8 kts.

ALISS
The Advanced Lightweight Influence Sweep System (ALISS) is currently under development by Alliant TechSystems to meet the US Navy's need for a lightweight acoustic and magnetic minesweeping system that can clear a beach assault area in significantly less time than is currently possible. The acoustic subsystem generates high-power, low-frequency sound waves to detonate acoustic influence mines. It will replace fixed frequency mechanical noisemakers used in the past with one that is software configurable to the required frequency band. The acoustic system will be integrated with the magnetic system.

7.5 Market Prospects
Although the major emphasis in recent years has been on minehunting and the destruction of mines using underwater vehicles, nevertheless, minesweeping should continue to play an important part in any MCM operation. While ground influence mines continue to proliferate and arsenals of such weapons expand, moored mines, – both influence and contact – will continue to be a major factor in any mine threat, as has been experienced in recent years in the Gulf. Hence the need for any navy which may consider it may have to face a mine threat to have within its mine countermeasures inventory the capability to conduct both wire and influence sweep operations.

Minesweepers continue to form the bulk of most MCM force inventories – with a few notable exceptions where minehunters predominate. Even in these cases, however, a minesweeping capability has not been neglected. Requirements for both wire sweep equipment and influence sweeps are not likely, therefore, to diminish. On the contrary, they may well expand world-wide over the next decade. Compared to dedicated minehunting systems, sweep systems are relatively inexpensive to purchase, although the need outlined above for systems with greater capability incorporating a measure of computer control has led to an increase in the cost of acquisition of such systems.

Minesweeping systems are also generally considered suitable for installation in COOP. Modifications to vessels, always supposing the correct type has been selected as a COOP for minesweeping operations, are relatively straightforward and can be accomplished fairly rapidly. As MCM force multipliers, therefore, the COOP with a modern minesweeping system installed is a fairly inexpensive but reasonably effective way in which to counter the threat of moored mines.

Apart from the need to acquire systems for new construction, there will always be an ongoing requirement for refurbishment and replacement of damaged systems currently in use, as well as the need to replenish stocks of expendable items such as cutters, explosive bolts and so on. Handling gear, and in particular winches, are subject to immense wear and tear due to the strains imposed upon them when handling sweep systems. These items of equipment need to be regularly refurbished and/or replaced. There is a continuous requirement too to replace damaged tow wires, sweep wires, magnetic loops and cables associated with handling acoustic generators.

The future requirement for minesweeping equipment, therefore, while not generally highly buoyant is nevertheless, a steady and ongoing one with any Navy which operates a min countermeasures force.

SECTION 8 - MACHINERY SYSTEM INVENTORIES

8.1 Introduction

Other than the minehunting system the machinery installation is probably one of the most important aspects of MCM design to be considered. This is because, of all the systems in an MCMV, the machinery is the one most likely to trigger an acoustic/magnetic influence mine.

Any machinery creates noise, and it is essential that every effort be made to reduce this self-generated noise as far as possible by the judicious and careful use of resiliently mounting any item of machinery, and then to mount the whole system on a suspended or resiliently mounted platform within the hull. One of the most significant areas relating to noise concerns the propellers, and very careful attention has to be paid to their design. Of interest in this area is the use of waterjets, cycloidal propellers and thrusters as opposed to the more conventional propeller. Machinery should, wherever possible, be acoustically treated to further reduce the noise signature. Finally, every care should be taken by machinery manufacturers to ensure that their systems run as silently as possible, which means careful design and use of materials in all moving parts, and in particular where those moving parts come into contact with one another for example gears, pistons and so on.

One way of reducing the noise signature is to mount the machinery above the waterline, but for reasons of weight (which affects stability) and the sheer size of some items of machinery, this is impractical. However, every effort should be made to mount as many items of machinery as is practical above the waterline, thus reducing as far as possible, without detriment to the stability of the vessel, the risk of unwanted noise.

One of the most significant areas relating to noise concerns the propellers, and very careful attention has to be paid to their design. Increasing use is now being made of waterjets, cycloidal propellers and thrusters as opposed to the more conventional propeller, principally because they exhibit a lower noise level.

Finally, MCMVs should be fitted with some form of noise monitoring equipment which can detect the slightest deviation in the accepted noise signature of the vessel, and aid the crew in identifying the cause of the problem. So sophisticated is the modern mine with its on board microprocessor, that ships which exhibit particular acoustic signatures can be recognised (assuming the enemy's intelligence gathering apparatus has provided him with such performance data) and mines programmed to react to that particular vessel.

Of equal importance to achieving a 'stealthy' MCMV design is the magnetic signature. This too is very largely determined by the materials used in the machinery. Even if an engine is built completely of non-magnetic materials, the manufacturing processes will still give rise to some inherent residual magnetism. This will vary in strength and direction, and during engine operation the permanent magnetism will change, and this, together with the induced magnetism resulting from the ship's geographical movements, will cause an unstable overall magnetic field.

On the other hand, completely ferromagnetic engines comprise more than 95% by weight of ferromagnetic materials. Such engines contain both indefinable and permanent amounts of magnetism which continually fluctuate while the engines are running and as the vessel moves about in the earth's magnetic field. The vessel thus exhibits a very strong, changing magnetic field.

The ship's degaussing system cannot be used to cancel out this magnetic field because the physical dimensions of the degaussing loops required to compensate for it are much greater than the size of the engine. To overcome the local field produced by the engine alone, it is necessary to surround it with degaussing loops. With these loops in all three planes the engine's magnetic field can be compensated only for the instantaneous magnetic state at the time of measurement. Any change in permanent magnetisation will result in an uncontrollable magnetic field. Furthermore, the alternating magnetic fields created by rotating ferromagnetic components cannot be compensated for by using degaussing loops.

In meeting the demands of modern MCMVs, manufacturers have adopted the approach of selecting diesels from a standard range, and wherever possible using demagnetising methods of construction and the use of non–ferromagnetic materials, and installing the engine onboard with controllable compensation techniques using degaussing coils.

While every vessel, irrespective of the materials used in its construction, will exhibit a certain residual magnetic signature, manufacturers must still devote a considerable effort to devising new methods of construction and the use of new materials in order to reduce the magnetic signature of their machinery as far as possible. Until recently traditional a–magnetic steels and conventional alloys have been used in the manufacture of propulsion machinery, principally because of the high stresses involved. The future, however, may well see a move towards the use of more exotic materials which exhibit high strain coefficients

Because of the nature of its task the MCMV does not operate at high speed, in fact the reverse, and so the diesel engine has become the standard propulsion unit. Auxiliary drive systems and power generation units, however, have used more unconventional systems, for example small gas turbines, electric drive and so on.

8.2 Europe

GERMANY

MTU

From experience gained with partially non–magnetic engines, MTU has developed a method of compensating standard ferrous engines. This has enabled the Company to offer its standard range of diesels for all applications, even where very low magnetic signature is required. Three methods of compensation have been developed to produce low magnetic signature diesels from the standard range.

The ADK (anti–dipole) compensation method achieves a stable reduction in ac and dc magnetic fields of a partly ferromagnetic engine by demagnetising rotating parts and large components (especially horizontal ones) and using a local longitudinal–acting automatic degaussing loop to compensate for the induced magnetic field. Stable permanent magnets are used to compensate for induced fields in vertically installed components.

The MRK (Controlled Magnetic Compensation) method has been developed from the ADK to reduce the magnetic field of a completely ferromagnetic engine. In this method the crankcase is treated using very stable permanent magnets. The crankcase is then treated with an alternating magnetic field, so that areas not under the influence of the permanent magnets are themselves permanently magnetised in the direction of the earth's field. The permanent magnetism thus produced cannot be destroyed under normal operating conditions. In addition to treating the crankcase, all other major components are demagnetised using an ac field technique. Induced fields are neutralised by overcompensation of the permanently magnetised crankcase/permanent magnets. The engine's induced field is compensated by horizontal degaussing loops which are automatically adjusted.

An MRK treated engine is adjusted for operation either throughout the northern hemisphere, the southern hemisphere or equatorial regions, and additionally adjusted for specific areas of operation.

The MRK–W (Controlled Magnetic Compensation for Worldwide Operation) uses, in addition to the system used in the MRK method, a degaussing loop to compensate for the engine's vertical magnetisation. The degree of compensation required in both the orientation of the permanent magnets and the current in the loop is determined by a microprocessor using data from a masthead mounted magnetometer.

The most popular magnetically treated engine from the MTU range is the 396 series, available in 6, 8, 12 and 16–cylinder versions.

Voith–Schneider

Voith–Schneider have developed a cycloidal propeller system which is eminently suitable for MCMVs as it overcomes many of the disadvantages and limitations associated with other propeller systems. The cycloidal system does not rely on rudders, azimuth thrusters or fixed or cp propeller types or any other form of auxiliary drive, all of which to some degree introduce an element of noise into the system. The cycloidal propeller exhibits a very low noise and magnetic signature compared to other current systems, whilst achieving the maximum possible degree of manoeuvrability.

The special features of the propeller are its extremely low magnetic signature, achieved by using practically all non–magnetic materials, and worm gear drive rather than the conventional bevel drive. The worm gear is used to convert the rotation about a horizontal axis of the input shaft to the rotation of the propeller rotor about a vertical axis. By using a worm gear in which there is only sliding movement on the surface of the teeth, providing a very smooth contact aided by a continuous thin film of oil, noise has been almost totally eliminated. The system is also capable of withstanding very severe shock loads. The other major feature of the system is that due to its vertical mounting and rotation, its efficiency is virtually unaffected by the direction of flow of water around it, a very important factor when carrying out intricate manoeuvres during the classification of a target.

UNITED KINGDOM

In the UK GEC–Alsthom Diesels have devoted considerable effort to reducing the magnetic signature of machinery. Standard diesels are built using strict demagnetisation construction methods. The engines are fitted with controllable compensation techniques using degaussing coils onboard the ship to reduce any inherent magnetism to an acceptable level. In addition the engine has to be capable of meeting other requirements which are far less tolerant than for other warships, such as noise and shock factors, compactness, low load, maintenance requirements, and so on.

This approach has been adopted with the Valenta diesel, substituting nearly 50% of the magnetic material in the original design with non-magnetic materials, but without destroying the standard form of the engine. Even with the use of compensating coils, however, it has not proved possible to completely reduce the induced magnetism to zero. However, it has been very considerably reduced, a factor which has been much improved by computer modelling to determine the positioning of coils around the engine itself. Further compensation is achieved from the compartment and ship coils, the control of currents in the coils being linked into the control system of the ship's degaussing system.

The reduced magnetic signature Valenta installed in the UK Royal Navy's '*Sandown*' class incorporates a degaussing system, nonferrous or low magnetic components and is raft mounted with fluid couplings.

TABLE 8A MACHINERY SYSTEMS- MINESWEEPERS, DRONES ETC

Class	Diesel	Manufacturer	Country	Output	Auxiliary	Manufacturer	Country	Output	Operator
MINESWEEPERS									
K 8 (M/S boat)	2 x Type 3-D-6			300/200kW					Vietnam
MSB (M/S boat)	1 x 64 HN9	Gray Marine		165/123kW					Thailand
MSB 07 (M/S boat)	2 x 4ZV20M	Mitsubishi	Japan	480/353kW					Japan
MSC (new minesweepers)	2 x	Brons/Werkspoor	Netherlands	2,176/1.6MW					Belgium
MSC (river minesweepers)	2			870/640kW					Romania
MSC 268	2 x 8-268A	GM	USA	880/656kW					South Korea
MSC 268	2 x 8-268A	GM	USA	880/656kW					Pakistan
MSC 268 & 292	4 x 6-71	GM	USA	696'/519kW					Iran
MSC 289	4 x 6-71	GM	USA	696/519kW					South Korea
MSC 294	2 x 268A	GM	USA	1,760/1.3MW					Greece
M 15	2 x			320235kW					Sweden
M 301	2 x								Yugoslavia
PO 2 Type 501	1 x 3-D-12		Russia	300/220kW					Bulgaria
T 43	2 x 9-D-8	Kolomna	Russia	2,000/1.47MW					Albania
T 43 (China built)	2 x 9-D-8	PCR/Kolomna	China	2,000/1.47MW					Bangladesh
T 43 Type 010	2 x 9-D-8	PCR/Kolomna	China	2,000/1.47MW					China
T 43	2 x 9-D-8	Kolomna	Russia	2,000/1.47MW					Egypt
T 43	2 x 9-D-8	Kolomna	Russia	2,000/1.6MW					Indonesia
T 43 Type 254	2 x 9-D-8	Kolomna	Russia	2,000/1.47MW					Russia
T 43	2 x 9-D-8	Kolomna	Russia	2,000/1.47MW					Syria

Class	Diesel	Manufacturer	Country	Output	Auxiliary	Manufacturer	Country	Output	Operator
T 301	3 x 6–cyl			900/661kW					Albania
T 301	3 x 6–cyl			900/661kW					Romania
Adjutant & MSC 268	2 x 8–268A	GM	USA	880/656kW					Spain
Adjutant & MSC 268	2 x 8–268A	GM	USA	880/656kW					Taiwan
Adjutant, MSC 268	4 x 6–71	GM	USA	696/519kW					Turkey
Adjutant, MSC 294	2 x L 1616	Waukesha	USA	1,200/895kW					Turkey
Aggressive	4 x 8–268A	GM	USA	1,760/1.3MW					Belgium
Aggressive	4 x ID–1700	Packard	USA	2,280/1.7MW					Spain
Aggressive	4 x ID–1700	Packard/Waukesha	USA	2,280/1.7MW					Taiwan
Alta	2 x 12V 396 TE84	MTU	Germany	3,700/2.72MW	2[1] x Eureka	Kvaerner	Sweden		Norway
	2 x 8V 396 TE54	MTU	Germany	1,740/1.28MW					
Arko	2 x MB 12V 493 TZ60	MTU	Germany	1,360/1MW					Sweden
Baltika Type 1380	1 x 18/22	ChISP	Russia	300/220kW					Russia
Bluebird	2 x 8–268A	GM	USA	880/656kW					Denmark
Bluebird	2 x 8–268	GM	USA	880/656kW					Thailand
Cape	4 x Type 2490 8V			1,300/970kW					Iran
Cove	4 x 6–71	GM	USA	696/520kW					Turkey
Dokkum	2 x V64	Fijenoord/MAN	Netherlands	2,500/1.84MW					Netherlands
Frauenlob Type 394	2 x MB 12V 493 TY70	MTU	Germany	2,200/1.62MW					Germany
Gassten	1 x			460/338kW					Sweden
Gilloga	1 x			380/279kW					Sweden
Ham	2 x YHAXM	GEC–Alsthom	UK	1,100/821kW					Yugoslavia

Class	Diesel	Manufacturer	Country	Output	Auxiliary	Manufacturer	Country	Output	Operator
Hameln Type 343	2 x 16V 538 TB91	MTU	Germany	6,140/4.5MW					Germany
Hisingen	1 x			380/279kW					Sweden
Kiiski	2 x 611 CSMP	Valmet	Finland	340/250kW					Finland
Kondor II Type 89	2 x Type 40-DM	Kolomna	Russia	4,408/3.24MW					Indonesia
Kondor II Type 89.2	2 x Type 40-DM	Kolomna	Russia	4,408/3.24MW					Latvia
Kondor II	2 x Type 40-DM	Kolomna	Russia	4,408/3.24MW					Uruguay
Krogulec Type 206F	2 x A-230S	Fiat	Italy	3,750/2.76MW					Poland
Kuha	2 x MT-380M	Cummins	UK	600/448kW					Finland
Leniwka Type 410S	1 x 6AL20/24	Puck-Sulzer	Switzerland	570/420kW					Poland
Lienyun	1 x			400/294kW					Vietnam
Musca	2 x			4,800/3.5MW					Romania
Natya I Type 266M	2 x Type M 504		Russia	5,000/3.67MW					Ethiopia & Eritrea
Natya I Type 266M	2 x Type M 504		Russia	5,000/3.67 MW					India
Natya I Type 266ME	2 x Type M 504		Russia	5,000/3.67MW					Lybia
Natya I/II Type 266M/DM	2 x Type M 504		Russia	5,000/3.67MW					Russia
Natya I Type 266M	2 x Type 504		Russia	5,000/3.67MW					Syria
Natya I Type 266ME	2 x Type M 504		Russia	5,000/3.67MW					Yemen
Nestin (river)	2 x Torpedo 12-cyl		Russia	520/382kW					Hungary
Nestin (river)	2 x Torpedo 12-cyl		Russia	520/382kW					Iraq
Nestin (river)	2 x Torpedo 12-cyl		Russia	520/382kW					Yugoslavia
Notec Type 207P	2 x M 401A1		Russia	1,874/1.38MW					Poland
Olya Type 1259	2 x Type 3D 6S11/235		Russia	471/346kW					Bulgaria

Class	Diesel	Manufacturer	Country	Output	Auxiliary	Manufacturer	Country	Output	Operator
Olya Type 1259	2 x Type 3D 6S11/235		Russia	471/364kW					Russia
River	2 x 6RKC	Ruston	UK	3,100/2.3MW					Bangladesh
River	2 x 6RKC	Ruston	UK	3,100/2.3MW					UK
Schutze	4 x	MTU/Maybach	Germany	4,500/3.3MW					Brazil
Sonya Type 1265	2 x Type 9-D-8	Kolomna	Russia	2,000/1.47MW					Bulgaria
Sonya Type 1265	2 x Type 9-D-8	Kolomna	Russia	2,000/1.47MW					Cuba
Sonya Type 1265	2 x Type 9-D-8	Kolomna	Russia	2,000/1.47MW					Ethiopia & Eritrea
Sonya Type 1265	2 x Type 9-D-8	Kolomna	Russia	2,000/1.47MW					Syria
Sonya Type 1265	2 x Type 9-D-8	Kolomna	Russia	2,000/1.47MW					Vietnam
Styrso	2 x DSI 14	Saab Scania	Sweden	1,104/812kW					Sweden
Ton	2 x Deltic 18A-7A	GEC-Alsthom	UK	3,000/2.24MW					South Africa
Vanya Type 257D	1 x Type M 503		Russia	2,502/1.84MW					Bulgaria
Vanya	1 x Type M 503		Russia	2,500/1.84MW					Syria
Vegesack	2 x MB	MTU	Germany	1,500/1.1MW					Turkey
Wosao	4 x M 50			4,400/3.23MW					China
Yevgenya Type 1258	2 x Type 3-D-12		Russia	600/440kW					Bulgaria
Yevgenya	2 x Type 3-D-12		Russia	600/440kW					India
Yevgenya	2 x Type 3-D-12		Russia	600/444kW					Syria
Yukto I/II	2 x								North Korea
Yurka Type 266	2 x Type M 503		Russia	5,350/3.91MW					Egypt
Yurka Type 266	2 x Type M 503		Russia	5,350/3.91 MW					Russia
Yurka Type 266	2 x Type M 503		Russia	5,350/3.91MW					Vietnam

Class	Diesel	Manufacturer	Country	Output	Auxiliary	Manufacturer	Country	Output	Operator
DRONES									
MSD	2 outboards	Yamaha	Japan	300/221kW					Australia
SAM	1 x TAMD70D	Volvo Penta	Sweden	210/154kW	1*				Sweden
SAV	1 x			350/257kW	1*				Denmark
Futi Type 312	1 x Type 12V 150C			300/220kW					China
Futi Type 312	1 x Type 12V 150C			300/220kW					Pakistan
Ilyusha Type 1253	2 x			500/367kW					Russia
Tanya Type 1300	1 x			270/					Russia
Troika	1 x D602	Deutz MWM	Germany	446/328kW					Germany
MISCELLANEOUS									
AN-2 (mine wareface/patrol)	2 x			220/162kW					Hungary
MCM (diving tenders)	2 x MGO 175 V16 ASHR	SACM	France	2,200/1.62MW	1†			70/51kW	France
MCS (training & support)	1 x	Skoda		2,176/1.6MW					Russia
MSA(T) (aux M/S tugs)	2 x	Stork Werkspoor	Netherlands	2,400/1.76MW					Australia
MSA(S) (Brolga)	1 x	Mirrlees	UK	540/403kW					Australia
MSA(S) (auxiliary M/S)	1 x D 346/V12 71	Caterpillar/GM	USA	480-359/358-264					Australia
MSA (aux minesweepers)	4 x Polar SF 16RS	Wartsila Nohab	Finland	4,600/3.38MW	1†	Gil Jet	Canada	575/429kW	Canada
MSI (Swiftships route survey)	2 x 12V 183 TA 61	MTU	Germany	928/682kW	1†			60/44kW	Egypt
MSR (M/S launches)	1 x			60/44kW					Greece
MST/ML (M/S support)	2 x			19,800/14.55MW					Japan

Class	Diesel	Manufacturer	Country	Output	Auxiliary	Manufacturer	Country	Output	Operator
MST (support ship)	2 x	MTU	Germany	1,310/963kW					Thailand
YAG (minehunting tenders)	2 x			2,000/1.47MW					Turkey
YDT (diving tenders)	2 x	GM	USA	165/123kW					Canada
YDT (diving tenders)	2 x	GM	USA	228/170kW					Canada
Type 742 (diver support)	2 x 160-cyl	Maybach MTU	Germany	5,200/3.82MW					Germany
Antares (route survey)	1 x 12P15-2SR	Baudouin	France	800/590kW	1†				France
Cosar (support ship/M/L)	2 x			400/4.7MW					Romania
Ejdern (sonobuoy craft)	2 x MD122	Volvo Penta	Sweden	366/269kW					Sweden
Fukue (support ship)	2 x 12ZC	Mitsubishi	Japan	1,440/1.06MW					Japan
Hayase Class (MST/ML)	4 x V6V22/30ATL	Kawasaki-MAN	Japan	6,400/4.7MW					Japan
Souya Class (MST/ML)	4 x V 22/30 ATL	Kauasaki-MAN	Japan	6,400/4.7MW					Japan

* Schottel pump jet propulsion
§ Water jets
¶ 2 x GM 8-268A diesels 880/656 in Shahrokh
† Thruster units

TABLE 8B MACHINERY SYSTEMS – MINEHUNTERS & MINEHUNTER/SWEEPERS

Class	Diesel	Manufacturer	Country	Output	Auxiliary	Manufacturer	Country	Output	Operator
MINEHUNTERS									
CME	2 x 6V 396 TB83	MTU–Bazan	Germany	1,523/1.12MW	2†			150/110k	Spain
MHC (Swiftships type)	2 x 12V 183 TE61	MTU		1,068/786kW	2*	Schottel	Germany		Egypt
					1†	White Gill	USA	300/224kW	
Bay	2 x 520–V8–S2	Poyaud	France	650/478kW	2*	Schottel			Australia
Circe	1 x	MTU	Germany	1,800/1.32MW					France
Flyvefisken	1 x LM 500¥	GE	USA	5,450/4.1MW	1 x 12V71	GM	USA	500/375k	Denmark
	2 x 16V 396 TB94	MTU	Germany	5,800/4.26MW	1†				
Frankenthal Type 332	2 x 16V 396 TB84	MTU	Germany	5,550/4.08MW	1				Germany
Gorya Type 1260	2 x			5,000/3.7MW					Russia
Huon (Gaeta)	1 x	Fincantieri GMT	Italy	1,986/1.46MW	3			506/372k	Australia
					3	Isotta Fraschini	Italy	1,440/1,058 kW	
Lida Type 1259.2	3 x			900/690					Russia
Oksoy	2 x 12V 396 TE84	MTU	Germany	3,700/2.72MW	2µ Eureka	Kvaerner	Sweden		Norway
	2 x 8V 396 TE54	MTU	Germany	1,740/1.28MW					
Osprey	2 x ID 36 SS 8V AM	Isotta Fraschini	Italy	1,600/1.18MW	1†			180/132k	USA
					3 x D 36	Isotta Fraschini	Italy	984kW	
					2			360/265kW	
Sandown	2 x Valenta 6RP200E	GEC–Alsthom	UK	1,500/1.12MW	2*				Saudi Arabia

Class	Diesel	Manufacturer	Country	Output	Auxiliary	Manufacturer	Country	Output	Operator
Sandown	2 x Valenta 6RP200E	GEC-Alsthom	UK	1,500/1.12MW	2*				UK
Swallow	2 x	MTU	Germany	2,040/1.5MW	1†			102/75k	South Korea
Tripartite	1 x A-RUB 215W-12	Stork Wartsila	Netherlands	1,860/1.37MW	2			240/176k	Belgium
					2†			150/110kW	
Tripartite	1 x A-RUB 215X-12	Stork Wartsila	Netherlands	1,860/1.37MW	2			240/179k	France
					2†			150/110kW	
Tripartite	2 x 12V 396 TC82	MTU	Germany	2,610/1.9MW	2			240/176k	Indonesia
					2†			150/110kW	
Tripartite	1 x A-RUB 215X-12	Stork Wartsila	Netherlands	1,860/1.35MW	2			240/179k	Netherlands
					2†			150/110kW	
Tripartite	1 x A-RUB 215X-12	Stork Wartsila	Netherlands	1,860/1.37MW	2			240/179k	Pakistan
					2†			150/110kW	
Yevgenya Type 1258	2 x Type 3-D-12		Russia	600/440kW					Angola
Yevgenya Type 1258	2 x Type 3-D-12		Russia	600/440kW					Cuba
Yevgenya Type 1258	2 x Type 3-D-12		Russia	600/440kW					Russia
Yevgenya Type 1258	2 x Type 3-D-12		Russia	600/440kW					Vietnam
Yevgenya Type 1258	2 x Type 3-D-12		Russia	600/440kW					Yemen
MINEHUNTERS/SWEEPERS									
MSC 07	2 x			1,800/1.33MW					Japan

MSC 322	2 x L-1616	Waukesha	USA	1,200/895kW					Saudi Arabia
MWV 50	2 x 8V 396 TB93	MTU	Germany	2,180/1.6MW					Taiwan
Adjutant	2 x 8-268A	GM	USA	880/656					Greece
Avenger (MCM 1-2)	4 x L-1616	Waukesha	USA	2,400/1.76MW	2	Hansome	USA	400/294k	USA
					1	Omnithruster		350/257kW	
Avenger	4 x ID 36 SS 6V AM	Isotta Fraschini	Italy	2,400/1.76MW	2	Hansome	USA	400/294k	USA
					1	Omnithruster		350/257kW	
Bang Rachan	2 x 12V 396 TB83	MTU	Germany	3,120/2.3MW	1				Thailand
Gaeta	1 x BL 230.8 M	Fincantieri GMT	Italy	1,985/1.46MW	3 ID 36	Isotta Fraschini	Italy	1,481/1.1	Italy
					3†			506/372kW	
Hatsushima	2 x YV122C-15/20	Mitsubishi	Japan	1,440/1.06MW					Japan
Hunt	2 x 9-59K Deltic	GEC-Alsthom	UK	1,900/1.42MW	1 x 9-55B	Ruston-Paxman	UK	780/582k	UK
					1†				
Kingston	4 x UD 23V12‡	Wartsila	Finland	3,000/2.2MW	2	Jeumont	France		Canada
					1†				
Landsort	4 x DSI 14	Saab Scania	Sweden	1,592/1.17MW					Singapore
Landsort	4 x DSI 14	Saab Scania	Sweden	1,592/1.17MW					Sweden
Lerici	1 x BL 230.8 M	Fincantieri GMT	Italy	1,985/1.46MW	3 ID 36	Isotta Fraschini	Italy	1,481/1.1	Italy
					3†			506/372kW	
Lerici	2 x 12V 396 TC82	MTU	Germany	2,605/1.91MW	3 ID 36	Isotta Fraschini	Italy	1,481/1.1	Malaysia
					2†				
Lerici	2 x 12V 396 TB83	MTU	Germany	3,120/2.3MW					Nigeria
Lindau Type 331	2 x MD 16V 538 TB90	MTU	Germany	4,000/2.94MW					Germany

Class	Diesel	Manufacturer	Country	Output	Auxiliary	Manufacturer	Country	Output	Operator
Lindau Type 351	2 x MD 16V 538 TB90	MTU	Germany	5,000/3.68MW					Germany
Notec II Type 207M	2 x M 401A			1,874/1.38MW	2			816/60k	Poland
River	2 x 12V 652 TB81	MTU	Germany	4,515/3.32MW					South Africa
Sonya Type 1265/1265M	2 x Type 9-D-8	Kolomna	Russia	2,000/1.47MW					Russia
Ton	2 x JVSS-12 Deltic	GEC-Alsthom	UK	3,000/2.24MW					Argentina
Uwajima	2 x 6NMU-TAI	Mitsubishi	Japan	1,400/1.03MW					Japan
Vanya Type 257D/DM	1 x M 503		Russia	2,502/1.84MW					Russia
Vukov Klanac (Ton)	2 x PA1 175	SEMT-Pielstick	France	1,620/1.19MW					Yugoslavia
Yaeyama	2 x 6NMU-TAI	Mitsubishi	Japan	2,400/1.76MW	1†			350/257k	Japan

† Thruster unit
* Schottel pump jet propulsion
μ Water jet
¥ Gas turbine
§ Diesel electric drive

TABLE 8C MAIN PROPULSION DIESEL ENGINES IN SERVICE AND ON ORDER*

Name	Builder	Navy	No x model	Output
Brons Werkspoor (Netherlands)				
4 x MSC	Belgium	Belgium	8 x	8,704
TOTALS 4 Units	**1 Country**	**1 Navy**	**8 Units**	**8,704**
Cummins (USA)				
6 x Kuha	Finland	Finland	12 x MT-380M	3,600
TOTALS 6 Units	**1 Country**	**1 Navy**	**12 Units**	**3,600**
Fiat (Italy)				
3 x Krogulec Type 206F	Poland	Poland	6 x A-230S	11,250
TOTALS 3 Units	**1 Country**	**1 Navy**	**6 Units**	**11,250**
Fijenoord/MAN (Netherlands)				
2 x Dokkum	Netherlands	Netherlands	4 x V64	5,000
TOTALS 2 Units	**1 Country**	**1 Navy**	**4 Units**	**5,000**
Fincantieri GMT (Italy)				
8 x Gaeta	Italy	Italy	8 x BL 230.8 M	15,880
6 x Huon	Australia	Australia	6 x	11,916
4 x Lerici	Italy	Italy	4 x BL 230.8 M	7,940
TOTALS 18 Units	**2 Countries**	**2 Navies**	**18 Units**	**35,736**
GEC-Alsthom Diesels (UK)				
2 x Ham	Yugoslavia	Yugoslavia	4 x YHAXM	2,200
13 x Hunt	UK	UK	26 x 9-59K Deltic	24,700
3 x Sandown	UK	Saudia Arabia	6 x Valenta 6RP200E	4,500
12 x Sandown	UK	UK	24 x Valenta 6RP200E	18,000
6 x Ton	UK	Argentina	12 x JVSS-12 Deltic	18,000
4 x Ton	UK	South Africa	8 x 18A-7A Deltic	12,000
TOTALS 40 Units	**2 Countries**	**5 Navies**	**80 Units**	**79,400**
General Motors (USA)				
3 x MSC 268	USA	South Korea	6 x 8-268A	2,640
2 x MSC 268	USA	Pakistan	4 x 8-268A	1,760
2 x MSC 268 & 292	USA	Iran	8 x 6-71	1,392
1 x MSC 268	USA	Iran	2 x 8-268A	880
5 x MSC 289	USA	South Korea	20 x 6-71	3,480
9 x MSC 294	USA	Greece	18 x 268A	15,840
8 x Adjutant & MSC 268	USA	Spain	16 x 8-268A	7,040
5 x Adjutant & MSC 268	USA	Taiwan	10 x 8-268A	4,400

Name	Builder	Navy	No x model	Output
11 x Adjutant & MSC 268	USA	Turkey	44 x 6–71	7,656
6 x Adjutant	USA	Greece	12 x 8–268A	5,280
2 x Aggressive	USA	Belgium	8 x 8–268A	3,520
1 x Bluebird	USA	Denmark	2 x 8–268A	880
2 x Bluebird	USA	Thailand	4 x 8–268A	1,760
4 x Cove	USA	Turkey	16 x 6–71	2,784
TOTALS 61 Units	**1 Country**	**10 Navies**	**170 Units**	**59,312**
Gray Marine (USA)				
12 x MSB	USA	Thailand	12 x 64 HN9	1,980
TOTALS 12 Units	**1 Country**	**1 Navy**	**12 Units**	**1,980**
Isotta Fraschini (Italy)				
12 x Avenger	USA	USA	48 x ID 36 SS 6V AM	28,800
12 x Osprey	USA	USA	24 x ID 36 SS 8V AM	19,200
TOTALS 24 Units	**1 Country**	**1 Navy**	**72 Units**	**48,000**
Kolomna (Russia)				
34 x T 43 Type 010	China	China	68 x Type 9–D–8*	68,000
4 x T 43	China	Bangladesh	8 x Type 9–D–8*	8,000
9 x Kondor II Type 89	East Germany	Indonesia	18 x Type 40–DM	39,672
2 x Kondor II Type 89	East Germany	Latvia	4 x Type 40–DM	8,816
4 x Kondor II Type 89	East Germany	Uruguay	8 x Type 40–DM	17,632
1 x T 43	Russia	Albania	2 x Type 9–D–8	2,000
2 x Yevgenya	Russia	Angola	4 x Type 3–D–12	1,200
2 x PO 2 Type 501	Russia	Bulgaria	2 x Type 3–D–12	600
4 x Sonya Type 1265	Rusia	Bulgaria	8 x Type 9–D–8	8,000
4 x Yevgenya	Russia	Bulgaria	8 x Type 3–D–12	2,400
6 x Olya Type 1259	Russia	Bulgaria	12 x Type 3D 6S11/235	2,826
4 x Sonya Type 1265	Russia	Cuba	8 x Type 9–D–8	8,000
12 x Yevgenya	Russia	Cuba	24 x Type 3–D–12	7,200
6 x T 43	Russia	Egypt	12 x Type 9–D–8	12,000
1 x Sonya Type 1265	Russia	Ethiopia	2 x Type 9–D–8	2,000
6 x Yevgenya	Russia	India	12 x Type 3–D–12	3,600
2 x T 43	Russia	Indonesia	4 x Type 9–D–8	4,000
5 x T 43 Type 254	Russia	Russia	10 x Type 9–D–8	10,000
3 x Olya Type 1259	Russia	Russia	6 x Type 3D 6S11/235	1,413
62 x Sonya Type 1265/1265N	Kolomna	Russia	124 x Type 9–D–8	24,000
27 x Yevgenya	Russia	Russia	54 x Type 3–D–12	16,200
1 x T 43	Russia	Syria	2 x Type 9–D–8	2,000
1 x Sonya Type 1265	Russia	Syria	2 x Type 9–D–8	2,000
5 x Yevgenya	Russia	Syria	10 x Type 3–D–12	3,000
5 x K 8	Russia	Vietnam	10 x Type 3–D–6	1,500

Name	Builder	Navy	No x model	Output
4 x Sonya Type 1265	Russia	Vietnam	8 x Type 9-D-8	8,000
2 x Yevgenya	Russia	Vietnam	4 x Type 3-D-12	1,200
5 x Yevgenya	Russia	Yemen	10 x Type 3-D-12	3,000
*Built under licence in China				
TOTALS 223 Units	**3 Countries**	**16 Navies**	**444 Units**	**368,259**
MTU (Germany)				
4 x Flyvefisken	Denmark	Denmark	8 x 16V 396 TB94	23,200
5 x Circe	France	France	5 x	9,000
5 x Frauenlob Type 394	Germany	Germany	10 x MB 12V 493 TY70	11,000
10 x Hameln	Germany	Germany	20 x 16V 538 TB91	6,140
6 x Schutze	Germany	Brazil	24 x	27,000
6 x Vegesack	Germany	Turkey	12 x MB	9,000
4 x MWV 50	Germany	Taiwan	8 x 8V 396 TB93	8,720
2 x Bang Rachan	Germany	Thailand	4 x 12V 396 TB83	6,240
10 x Frankenthal	Germany	Germany	20 x 16V 396 TB84	55,550
8 x Lindau Type 331	Germany	Germany	16 x MD 16V 538 TB90	32,000
6 x Lindau Type 351	Germany	Germany	12 x MD 16V 538 TB90	30,000
4 x Lerici	Italy	Malaysia	8 x 12V 396 TC82	10,420
2 x Lerici	Italy	Nigeria	4 x 12V 396 TB83	6,240
2 x Tripartite	Netherlands	Indonesia	4 x 12V 396 TC82	5,220
9 x Alta/Oksoy	Norway	Norway	18 x 12V 396 TE84	33,300
9 x Alta/Oksoy	Norway	Norway	18 x 8V 396 TE54	15,660
4 x River	South Africa	South Africa	8 x 12V 652 TB81	18,060
6 x Swallow	South Korea	South Korea	12 x	12,240
4 x CME	Spain	Spain	8 x 6V 396 TB83*	6,092
1 x Arko	Sweden	Sweden	2 x MB 12V 493 TZ60	1,360
3 x MHC	USA	Egypt	6 x 12V 183 TE61	3,204
*Built under licence in Spain				
TOTALS 110 Units	**11 Countries**	**16 Navies**	**227 Units**	**384,856**
Mitsubishi (Japan)				
2 x MSB 07	Japan	Japan	4 x 4ZV20M	960
20 x Hatsushima	Japan	Japan	40 x YV122C-15/20	28,800
9 x Uwajima	Japan	Japan	18 x 6NMU-TAI	12,600
3 x Yaeyama	Japan	Japan	6 x 6NMU-TAI	7,200
TOTALS 34 Units	**1 Country**	**1 Navy**	**68 Units**	**49,560**
Packard (USA)				
4 x Aggressive	USA	Spain	16 x ID-1700	9,120
4 x Aggressive	USA	Taiwan	16 x ID-1700	9,120

Name	Builder	Navy	No x model	Output
TOTALS 17 Units	**1 Country**	**2 Navies**	**32 Units**	**18,240**
Poyaud (France)				
2 x Bay	Australia	Australia	4 x 520–V8–S2	1,300
TOTALS 2 Units	**1 Country**	**1 Navy**	**4 Units**	**1,300**
Puck–Sulzer (Switzerland)				
2 x Leniwka Type 410S	Poland	Poland	2 x 6AL20/24*	1,140
*Built under licence in Poland				
TOTALS 2 Units	**1 Country**	**1 Navy**	**2 Units**	**1,140**
Russia				
4 x Vanya Type 257D	Russia	Bulgaia	4 x Type M 503	10,008
4 x Yurka Type 266	Russia	Egypt	8 x Type M 503	21,400
1 x Natya I Type 266M	Russia	Ethiopia	2 x Type M 504	5,000
12 x Natya I Type 266M	Russia	India	24 x Type M 504	60,000
8 x Natya I Type 266ME	Russia	Libya	16 x Type M 504	40,000
17 x Notec I/II	Russia	Poland	34 x M 401A1	31,858
1 x Baltika Type 1380	Russia	Russia	1 x 18/22	300
2 x Gorya Type 1260	Russia	Russia	4 x	20,000
24 x Lida Type 1259.2	Russia	Russia	24 x	21,600
27 x Natya I/II Type 266M/DM	Russia	Russia	54 x Type M 504	270,000
13 x Vanya Type 257D/DM	Russia	Russia	13 x Type M 503	32,526
2 x Yurka Type 266	Russia	Russia	4 x Type M 503	21,400
1 x Natya I Type 266M	Russia	Syria	2 x Type M 504	5,000
2 x Vanya Type 257D	Russia	Syria	2 x Type M 503	5,004
2 x Yurka Type 266	Russia	Vietnam	4 x Type M 503	21,400
1 x Natya I Type 266M	Russia	Yemen	2 x Type M 504	10,000
TOTALS 121 Units	**1 Country**	**10 Navies**	**198 Units**	**575,496**
Ruston (UK)				
4 x River	UK	Bangladesh	8 x 6RKC	12,400
5 x River	UK	UK	10 x 6RKC	15,500
TOTALS 9 Units	**1 Country**	**2 Navies**	**18 Units**	**27,900**
Saab Scania (Sweden)				
4 x Landsort	Sweden	Singapore	16 x DSI 14	6,368
4 x Styrso	Sweden	Sweden	8 x DSI 14	4,416
7 x Landsort	Sweden	Sweden	28 x DSI 14	11,144
TOTALS 15 Units	**1 Country**	**2 Navies**	**52 Units**	**21,928**
SEMT–Pielstick (France)				
2 x Vukov Klanac	France	Yugoslavia	4 x PA1 175	3,240
TOTALS 2 Units	**1 Country**	**1 Navy**	**4 Units**	**3,240**
Stork Wartsila (Netherlands)				
7 x Tripartite	Belgium	Belgium	7 x A–RUB 215W–12	13,020

Name	Builder	Navy	No x model	Output
10 x Tripartite	France	France	10 x A-RUB 215W-12	18,600
15 x Tripartite	Netherlands	Netherlands	15 x A-RUB 215W-12	27,900
3 x Tripartite	France	Pakistan	3 x A-RUB 215W-12	5,580
TOTALS 35 Units	**3 Countries**	**4 Navies**	**35 Units**	**65,100**
Valmet (Finland)				
7 x Kiiski	Finland	Finland	14 x 611 CSMP	2,380
TOTALS 7 Units	**1 Country**	**1 Navy**	**14 Units**	**2,380**
Wartsila (Finland)				
12 x Kingston	Canada	Canada	48 x UD 23V12	36,000
TOTALS 12 Units	**1 Country**	**1 Navy**	**48 Units**	**36,000**
Waukesha (USA)				
11 x Adjutant, MSC 294	USA	Turkey	22 x L 1616	13,200
4 x MSC 322	USA	Saudi Arabia	8 x L-1616	4,800
2 x Avenger (MCM1-2)	USA	USA	8 x L-1616	4,800
TOTALS 17 Units	**1 Country**	**3 Navies**	**38 Units**	**22,800**
Unknown				
2 x T 301		Albania	6 x	1,800
1 x Wosao		China	4 x M 50	4,400
6 x Nestin		Hungary	12 x Torpedo	3,120
2 x Cape		Iran	8 x Type 2490 8V	2,600
2 x Nestin		Iraq	4 x Torpedo	1,040
3 x MSC 07		Japan	6 x	5,400
25 x MSC		Romania	50 x	21,750
12 x T 301		Romania	36 x	10,800
4 x Musca		Romania	8 x	19,200
5 x M 15		Sweden	10 x	1,600
3 x Gassten		Sweden	3 x	1,380
3 x Gilloga		Sweden	3 x	1,140
4 x Hisingen		Sweden	4 x	1,520
2 x Lienyun		Vietnam	2 x	800
9 x Nestin		Yugoslavia	18 x Torpedo	4,680
TOTALS 110 Units		**11 Navies**	**174 Units**	**81,230**

*Main Propulsion Systems for minesweepers and minehunters - Drones and miscellaneous auxiliary vessels not included

TABLE 8D MAIN PROPULSION DIESEL ENGINE SALES*

Manufacturer	Units Installed	Output	No of User Navies
Kolomna	444	368,259	16
MTU	227	384,856	16
Russia	198	575,496	10
General Motors	170	59,312	10
Paxman	80	79,400	5
Isotta Fraschini	72	48,000	1
Mitsubishi	68	49,560	1
Saab Scania	52	21,928	2
Wartsila	48	36,000	1
Stork Wartsila	35	65,100	4
Waukesha	38	22,800	3
Packard	32	18,240	2
Fincantieri GMT	18	35,736	2
Ruston	18	27,900	2
Valmet	14	2,380	1
Cummins	12	3,600	1
Gray Marine	12	1,980	1
Brons Werkspoor	8	8,704	1
Fiat	6	11,250	1
Fijenoord/MAN	4	5,000	1
Poyaud	4	1,300	1
SEMT–Pielstick	4	3,240	1
Puck–Sulzer	2	1,140	1
Unknown	174	81,230	11

* Main Propulsion Systems for minesweepers and minehunters – Drones and miscellaneous auxiliary vessels not included

UK Hunt class coastal minesweepers/minehunters HMS Chiddingfold (M37) and HMS Quorn *(H M Steele)*

8.3 Market Prospects

Without doubt the world market in main propulsion diesel engines for MCMVs has been well and truly captured by MTU of Germany. In total, some 227 engines with a total output of 384,856 horsepower are currently in service or on order for 16 navies around the world.

The only other supplier to exceed this total is Russia with a total of 642 engines with a total output power of 943,755 horsepower. However, the majority of these engines were supplied to units which were built for the former Soviet Navy and later transferred to other friendly navies, or built in small numbers directly for friendly navies – a captive market for at the time many of those navies were under Soviet influence. Furthermore, the majority of these units are now nearing the end of their life, and under present circumstances it seems unlikely that many of them will be replaced. Currently there is not a large order book for Russian MCMVs. The domestic market is virtually at a standstill, and of the navies formerly tied to Soviet military hardware, very few will be in a position to replace obsolete units. Finally, it seems unlikely that Russia will have very much of an impact on the open market with her MCMV designs, and hence there will be virtually no requirement for Russian diesel engines.

On the other hand, having secured the majority of the export market with its diesel engines, it does not seem likely that MTU will lose this position in the near future. On the contrary, there is every likelihood that it will continue to secure the majority of future orders for main propulsion systems for MCMVs.

The only other major contender currently providing diesels for this market is Paxman (a member of the GEC–Alsthom Group) of the UK. At present there are some 80 Paxman diesels with a total output of 79,400 horsepower in service in 5 navies around the world. However, it should be noted that some of these diesels are now very old, being installed in vessels that are nearing the end of their life. Nevertheless, Paxman have achieved success in the export market and will probably continue to achieve some overseas sales.

Table 10C also indicates that General Motors of the USA is a leading contender for the supply of main propulsion diesels for mine warfare vessels. Again, however, this is misleading for the bulk of the 170 diesels manufactured were for vessels originally built for the USN and later transferred to overseas navies. The bulk of these units too are now nearing the end of their life, and General Motors has not supplied diesels to new MCMV construction for many years.

In Italy two manufacturers have achieved a measure of success with their diesels – Fincantieri GMT and Isotta Fraschini. Fincantieri GMT has only supplied diesels for vessels built in Italy, and any future sales would seem to be dependent on Italy winning orders for more indigenous MCMV designs based on the *'Lerici/Gaeta'*. However, not all the *'Lerici'* designs built for export have been fitted with Fincantieri GMT diesels, and Malaysia and Nigeria both opted to fit MTU diesels in their vessels.

Isotta Fraschini achieved a notable success with its diesels, winning the order to supply propulsion power to both the new mine warfare vessel designs of the US Navy. The first class, the *'Avenger'*, were originally to have been powered by Waukesha diesels, but following problems with the first two units, the remainder of the class was fitted with Isotta Fraschini diesels, which were also selected for the *'Osprey'* design, itself based on the Italian *'Lerici'*.

In Scandinavia Wartsila and Stork Wartsila between them have supplied 83 diesels with a total output of 101,100 horsepower. Wartsila are supplying 48 diesels with an output of 36,000 horsepower for the 12 *'Kingston'* class vessels of the Canadian MCDV Programme, while Stork Wartsila have supplied the main diesel propulsion system for the extensive *'Tripartite'* programme.

Also in Scandinavia, Saab Scania have supplied the main propulsion system for the *'Landsort'* class – which has also been sold to Singapore. They are also supplying the main propulsion system for the new *'Styrso'* class currently under construction.

In the Far East Mitsubishi supply solely to vessels built for the JMSDF. It is unlikely in the foreseeable future that the Company will be able to penetrate the export market.

The future market for main propulsion diesel engines will largely depend on the new build market. Diesels are very rugged, and if treated with reasonable care should last the lifetime of the hull. Other than sales for new construction, the main activity in this sector will be the refurbishment of units already in service. As regards new construction, MTU will probably secure the bulk of any new orders, with Paxman and possibly Saab Scania and Wartsila providing the main competition.

The Danish Flyvefisken class patrol craft Havkatten (left) and Hajen fitted as minehunters
(H M Steele)

SECTION 9 - MCMV FORCES - CURRENT & FUTURE PROGRAMMES 1996-2005

9.1 Introduction

Mine countermeasures vessel construction over the last two decades has remained at a relatively steady rate worldwide. Not surprisingly the majority of new construction has been undertaken in Europe and Scandinavia, a region where the effect of mine warfare during two World Wars had an almost crippling effect on some of the protagonists, and caused major disruption to the movement of shipping for some of the others.

More recently the strategic and tactical impact that mine warfare can have has been seen in the Middle and Far East Regions, and this too has had an effect in the development of mine warfare forces in these regions.

However, it must be noted that with the ending of the Cold War and its effect worldwide, there seems to have been a noted decline in the willingness of some navies to devote the necessary funds to the maintenance of a modern MCM capability.

The one region which has paid virtually no attention to the need for adequate MCM forces has been the South America and Caribbean Region. Furthermore, there does not at present appear to be any major requirement forthcoming from this region for modern MCM forces.

Overall, while some navies continue to place considerable and growing importance on the need to maintain and build up adequate MCM forces to counter any potential future mine threat, others appear to see little need for such a capability. This is surprising when put into context against the development and expansion of many navies, and more particularly with the continuing development of submarine forces worldwide.

Any nation now possessing even just one or two modern high value surface warships as well as submarines must face the fact that in times of rising tension leading to open hostilities these units will be highly vulnerable to a mine threat unless a modern MCM capability is available. This, however, is not the only factor which should be considered. Many nations rely heavily on seaborne trade for their economic independence, and trade routes too will be highly vulnerable to a mine threat unless a modern MCM capability is at hand.

The Middle East is a region in particular where countries are rapidly building up their naval forces in a highly charged political atmosphere, but where very little effort is being paid to the need for an adequate MCM capability. Some countries are, it seems, content to leave any need for MCM capability to the readiness of Western navies to come to their aid – as has happened in the past. But the threat may materialise and have its effect before outside forces can come to the aid of the country under attack.

Europe will endeavour, in so far as economics and the reduced tension between East and West will allow, to maintain its current MCM capability. Many navies have recently completed, or are completing, major programmes of new construction and plans are in hand for major upgrade programmes to modernise older construction.

While it does not seem likely that there will be large programme for new construction in the near future, it is highly probable that there will be an ongoing requirement worldwide for small numbers of vessels – mainly required to replace obsolete hulls and maintain a credible MCM capability. Where funds are likely to be forthcoming some navies may begin to devote more of their resources to developing an MCM capability.

Overall, therefore, the picture would seem to show a relatively low rate of continual new construction, backed up by modernisation programmes.

Table 9.2 indicates those regions where, over the next 10 to 20 years, there is potential for new construction and modernisation. Assuming hull replacement on a one for one basis, world–wide there could be a requirement for about 359 hulls over the next 10 to 20 years to replace boats that are now over 20 years old. However, many of these are either units of the Russian Navy which at present does not appear to be in a position to either require or indeed to order replacement hulls, or else they are units transferred by Russia or the United States and which are either currently being replaced by new construction or are in navies which do not at present have funds available to replace obsolete units.

As far as modernisation and upgrading is concerned the total requirement will be for some 245 hulls, although a number of these have received a major upgrading or modernisation. The upgrade/retrofit market covers hulls that are at the present time between 11 and 20 years old. Bearing in mind the rapid advances being made in technology, particular in command systems and sonars, and the general development in MCM capability, some 92 hulls of between 6 to 10 years old will also need varying degrees of upgrading over the next five to 10 years.

TABLE 9.1 CURRENT NUMBERS OF MCMVs IN ORDER OF BATTLE

Region 1		Region 2		Region 3		Region 4		Region 5	
Country	**Number**	**Country**	**Number**	**Country**	**Number**	**Country**	**Number**	**Country**	**Number**
Albania	3	Bangladesh	5	Australia	10	Argentina	6	Angola	2
Belgium	9	Egypt	15	China	95	Brazil	6	Ethiopia	2
Bulgaria	20	India	18	Indonesia	13	Canada	12	Nigeria	2
Denmark	11§	Iran	5	Japan	37	Cuba	16	S. Africa	12
Finland	13	Iraq	2	N. Korea	23	Uruguay	4		
France	26	Libya	8	S. Korea	14	USA	22		
Germany	58	Pakistan	7	Malaysia	4				
Greece	19	Saudi Arabia	7	Singapore	4				
Hungary	51	Syria	10	Taiwan	13				
Italy	12	Yemen	6	Thailand	12				
Latvia	2			Vietnam	15				
Netherlands	17								
Norway	7								
Poland	22								
Romania	43								
Russia†	173								
Spain	12								
Sweden	32								
Turkey	29								
UK	23								
Yugoslavia	15								
TOTAL 21	**597**	**10**	**83**	**11**	**240**	**6**	**66**	**4**	**18**

* Figures in all tables are projected to the end of 1996. Figures include COOP (where known), river clearance vessels, drones, route survey vessels, MCM support vessels and tenders, but does not include minelaying vessels

§ Includes six MCM modules allocated for Danish *'Flyvefisken'* class

† The figures for units listed in the Order of Battle are open to question, but are considered to be representative. They do not include transfers overseas.

Region 1: Europe & Scandinavia
Region 2: Middle East, North Africa, Gulf & Western Asia
Region 3: Asia, Pacific & Australasia
Region 4: North America, South America & Caribbean
Region 5: Africa

TABLE 9.2 MCMV AGE 1972-96 WORLD MARKET FOR HULLS

Region	**1992-96**	**1987-91**	**1982-86**	**1977-81**	**1972-76**	**Pre-1972**	**TOTAL**
Age in Years	0-5	6-10	11-15	16-20	21-25	over 25	
Europe & Scandinavia	41	45	58	37	187§	111	**479**
Middle East, Gulf, North Africa & Western Asia	8	10	20	15	8	11	**72**
Asia, Pacific & Australasia	21	23	47	48§	9	17	**165**
South America & Caribbean	4	0	7	9	4	8	**32**
North America & Africa	18	14	0	4	0	4	**40**
GRAND TOTAL	**92**	**92**	**132**	**113**	**208**	**151**	**788**

§ A large number of these are Russian or Chinese units built over many years and actual dates are not readily available. The majority, however, are in need of replacement.

9.2 Europe and Scandinavia

In Europe and Scandinavia well over half the current MCMV strength (298 hulls) is over 20 years old. While MCMV hulls are rugged and in most navies have tended to have a life expectancy in the region of 30-35 years if carefully looked after, this is not a situation that can be expected to continue. The bulk of the 298 hulls are fairly simple minesweepers, which do have a very long life expectancy. Furthermore, the bulk of these units are either of Russian or American origin, which were built in large numbers in the early days of the Cold War, and subsequently transferred to friendly navies.

Many of these hulls are now being listed for disposal, and while some navies have embarked on replacement construction, others, including former Soviet satellite navies, are certainly in no position to replace their obsolete hulls. Financial stringencies are a very major factor in these circumstances, and usually any resources which are available are devoted towards maintaining large, high value units.

On this basis it is very possible that a number of these navies will meet any MCM requirement from existing mercantile hulls - probably trawlers - and convert them for minesweeping operations, purchasing the necessary equipment from abroad. It is most unlikely that any of the former Warsaw Pact navies will be in a position to acquire minehunters for some years.

Of the NATO navies in the region most have, in recent years, embarked on new construction to replace obsolete hulls. Norway and Spain are the most recent to embark on new programmes, while the UK has an ongoing programme of new construction. France abandoned a new programme of construction and is unlikely to embark on further new construction for some years. Germany is just completing a major programme of new construction to replace obsolete hulls. Belgium, is starting a new programme of construction, and may put some of its Tripartite hulls up for sale, as may the Netherlands.

Countries who are planning to modernise their forces with new construction are Greece, Portugal and Turkey. Russia has a large fleet of obsolete MCMVs, but such is the poor state of the economy that any programme of new construction in the near future is likely to suffer severe delays - if indeed any such programme is even approved.

As far as modernisation is concerned, the UK will, in the near future, embark on a major programme to upgrade the 13 *'Hunt'* class. Germany and the Netherlands have also formulated plans to modernise and upgrade their existing MCM capability. The French Tripartite will also be upgraded. By the end of the decade the Swedish *'Landsort'* class will be in need of upgrading and modernisation, while Finland will have to give consideration to its possible future needs in this area. The four Italian *'Lerici'* class will also be due for a mid life update within the next few years.

The one navy over which a question mark hangs is the Polish Navy. Although operating a number of relatively new build MCMVs, the equipment is not as capable as that in some of the NATO navies and consideration will have to be given to ways in which the Polish MCM capability can be upgraded between now and the turn of the century.

While many of the foregoing remarks concerning replacement relate to minesweepers, the introduction of minehunters into navies is a relatively new factor. Of these the British *'Hunt'* class is the oldest, and many navies will be keeping a close watch on this mid-life upgrade programme before embarking on the upgrading of their own minehunters. It is unlikely that minehunters will achieve quite the same life expectancy as minesweepers, but with careful monitoring, maintenance and upkeep, there is no reason why the hulls should not last at least for 25 if not 30 years. Much will depend on how easy and cost effective it proves to upgrade a small ship carrying a large amount of highly sophisticated electronic equipment.

TABLE 9.3 MCMV AGE 1970–96¥ EUROPE & SCANDINAVIA

Country	1992–96 0–5 years	1987–91 6–10 years	1982–86 11–15 years	1977–81 16–20 years	1972–76 21–25 years	Pre–1972 over 25 years	TOTAL
Albania						3	**3**
Belgium		5	2			2	**9**
Bulgaria	2	4	4†		4	6	**20**
Denmark						1	**1***
Finland			7		6		**13**
France	1	3	6		5‡		**15**
Germany	10	10				19µ	**39**
Greece	2†					13†	**15**
Hungary				6			**6**
Italy	8		4				**12**
Latvia	2†						**2**
Netherlands		4	11			2	**17**
Norway	7						**7**
Poland	4	8	7			3	**22**
Romania		4	<	25§	>	12	**41**
Russia		<			166§	>	**166**
Spain						12†	**12**
Sweden	2	2	4		3	13	**24**
Turkey						17†	**17**
UK	3	5	13	2			**23**
Yugoslavia				4	3	8	**15**
TOTAL	**41**	**45**	**58**	**37**	**187**	**111**	**479**

¥ Figures do not cover drones or miscellaneous craft listed in previous tables

* Plus a number of *'Flyvefisken'* class for which four minehunting modules are available

† Transferred – hulls may be considerably older

‡ To be decommissioned in 1997–98

µ Being replaced by *'Hameln'* and *'Frankenthal'* classes

§ Many of these were built over a large number of years and actual dates are not readily available. The majority, however, are in need of replacement.

9.3 Middle East, Gulf, North Africa & Western Asia

In this region the number of overage hulls considered to be in need of replacement totals 19. Countries need to consider replacement of overage hulls are: Egypt; Iran; Pakistan; and Syria. Other countries which received units transferred from the Soviet Union or America and which are now overage are Iraq and the Yemen.

In view of the international political situation it is most unlikely that Iraq will be able to replace its overage units. Of the other countries Egypt has fairly recently acquired three new MCMVs. However, the 10 overage units have not yet been withdrawn from service. Although there are no immediate plans to acquire more MCMVs, these 10 units will have to be replaced at some point in the relatively near future if the country is to retain a credible MCM force. Any such future acquisition could really only be obtained under FMS funding, and in this case it is possible that more Swiftships–type MCMVs may be obtained.

Although none of its units are overage, India does have plans to acquire 10 more MCMVs. This is necessary in view of the enormous length of its coastline bordering two major ocean areas (the Arabian Sea and the Bay of Bengal), numerous ports and shipping routes, and two major naval fleets – all of which are vulnerable to a mine threat. Whether India will again seek to acquire these units from Russia or build domestically, is as yet undecided. However, initial work has been undertaken in the Goa Shipyard with a view to the possibility of indigenous construction of GRP MCMVs. In this case, India will certainly seek licence construction of a European design. Much, however, will depend on the availability of funding for this project – which has already been in discussion for many years. Of greater interest will be any decision to upgrade the existing MCMV fleet, and this must now be a high priority for the Indian Navy. The major need is to acquire a modern minehunting system – and this may well have to be supplied by an overseas country as the Indian electronics industry is probably not yet in a position to supply the necessary sophisticated command system and minehunting sonars.

Iran is unlikely to be favoured with the supply of new MCMVs from the West. Any requirement to replace obsolete units will most likely be met by China with a T 43 new build design.

Libya, regarded by most countries as being an unacceptable customer – even by Russia – is unlikely to be in a position to either refurbish or replace its existing fleet, while Syria is not in a financial position to either upgrade or replace its existing fleet. The Yemen is in a similar position.

Pakistan has embarked on the acquisition of three new Tripartite minehunters from France and these will replace the obsolete units. There are no immediate plans to acquire more vessels, but future plans may include additional units to be acquired in the time span 2000–2005.

In the Gulf the only country to have recognised and accepted the need for a modern MCM capability is Saudi Arabia. Three modern British *'Sandown'* class minehunters are being acquired, and plans exist to acquire a further three units at some future date. These will replace the existing four MSC 322 vessels. As yet no other Gulf Navy has definite pans to develop an MCM capability – although all are utterly dependent on maintaining shipping routes free of obstruction in the Gulf for the export of oil– and many are building sizeable naval forces with high value units.

TABLE 9.4 MCMV AGE 1972–96[¥]
MIDDLE EAST, GULF, NORTH AFRICA & WESTERN ASIA

Region/Country	1992–96 0–5 years	1987–91 6–10 years	1982–86 11–15 years	1977–81 16–20 years	1972–76 21–25 years	Pre–1972 over 25 years	TOTAL
Bangladesh			4†				4
Egypt	3				6†	4†	13
India		5	7	6			18
Iran						5†	5
Iraq				2†			2
Libya			6	2			8
Pakistan	2					2	4
Saudi Arabia	3			4			7
Syria			6†	1†	2†		9
Yemen		5†	1†				6
TOTAL	**8**	**10**	**24**	**15**	**8**	**11**	**76**

¥ Figures do not cover drones or miscellaneous craft listed in previous tables
† Transferred – hulls may be considerably older

9.4 Asia, Pacific and Australasia

Other than Europe, the Far East is the one region which has seen the greatest efforts to develop and build up a credible MCM capability to meet the modern mine threat.

Australia has embarked on a major programme of construction to acquire a new class of minehunters based on the Italian *'Gaeta'* design. In addition the experience of trials over a number of years with the COOP concept has enabled Australia to develop an effective influence sweep system which can be easily installed on such craft, and elements of this new system (AMASS) are already achieving export success. Once the programme of new construction is complete, the Australian defence industry will be in a powerful position to meet the future MCM needs of the region. However, Australia will have to compete hard with European countries who have already captured a leading share of the market in this region.

Altogether a total of about 26 hulls in this region will need replacing over the next five years if the current level of MCM strength is to be maintained. Countries who will need to address this question are: China; Indonesia; South Korea; Taiwan; Thailand; and Vietnam. Of these, all but China and Vietnam are known to have plans to acquire new units in the fairly near future. Doubtless even China has plans to upgrade its existing MCM capability which is either overage or obsolete in the face of the modern mine threat. With regard to future programmes, China will probably develop indigenous designs of both minesweeper and minehunter, but the technology for the latter will probably lag some way behind that available in the West.

Indonesia has, in the past, acquired minehunters from the Netherlands, and most recently units from the former East German Navy. These latter were greatly in need of major upgrading when acquired, to enable them to counter the modern influence mine. To meet this requirement elements of the Australian AMASS system are being procured. As for future acquisitions, which are restricted by the limited availability of funds, Indonesia may well seek to acquire hulls from Australia, for which the probable timescale for both countries would be eminently suitable. Any future requirement will, however, be hotly contested by the major European suppliers from France, Germany, Italy, Sweden and the UK. On the other hand Indonesia may well be content to acquire more second–hand units of the Tripartite type of which examples are already operational in the Indonesian fleet. Again the timing could be just right for this acquisition by Indonesia, for both Belgium and the Netherlands are understood to be willing to put some of their own units now surplus to requirements, up for sale.

Japan will continue to acquire units domestically to maintain a modern force of about 20 units. A new class of MCMV (MSC 07) is under construction, being fitted with European minehunting equipment – a major departure from previous Japanese experience where all such equipment was supplied by the USA. It is believed that the Japanese are comparing European and American minehunting equipments with a view to deciding on their future source of supply for new designs which may be procured around the turn of the century. The new designs may well combine minehunting and sweeping, and two classes may be needed to meet a requirement for a coastal vessel and an ocean going MCMV. The designs may be based on the *'Yaeyama'* and MSC 07, of which at present only three examples of each have, or are being procured.

South Korea has already indicated that it intends to acquire a new, larger class of MCMV, which will almost certainly be built by the Kang Nam Corporation which built the existing *'Swallow'* class. While the design will be developed in South Korea, the command system and sonars will undoubtedly be purchased from Europe. The United Kingdom will be the strongest contender to supply the electronics systems, but the bidding will be hotly contested by other European countries.

Taiwan also has a requirement for additional MCMVs, both to replace overage units and build up its MCMV strength. Sourcing, however, as with all Taiwanese military projects, will prove a major problem.

Thailand has a major requirement for new MCMVs – both to replace overage hulls and build up its existing capability. With the acquisition of its carrier from Spain, the latter country must now be considered to be a major contender for any future MCMV programme, especially as Spain is embarking on the construction of a new GRP–hulled MCMV for its own Navy.

In the upgrade and modernisation market, Malaysia will soon have to give serious consideration to a major upgrading of its four *'Lerici'* class minehunters – a project which will need to be implemented in the time span 2000–2005.

Of the various regions, this is certainly the one where the bulk of the market for new construction MCMVs is likely to be over the next 10 years. A growing market for upgrading and modernisation also exists in the region.

TABLE 9.5 MCMV AGE 1972–96[¥]
ASIA, PACIFIC & AUSTRALASIA

Region/Country	1992–96 0–5 years	1987–91 6–10 years	1982–86 11–15 years	1977–81 16–20 years	1972–76 21–25 years	Pre–1972 over 25 years	TOTAL
Australia		1	1				**2**
China		<		35[§]		>	**35**
Indonesia		2			7[†]	4[†]	**13**
Japan	10	8	10	4			**32**
Korea (North)			23				**23**
Korea (South)	3	2	1		2	6	**14**
Malaysia			4				**4**
Singapore	4						**4**
Taiwan	4[†]	4				5	**13**
Thailand		2				2	**4**
Vietnam		4[†]	4[†]	9[†]			**17**
TOTAL	**21**	**23**	**43**	**48**	**9**	**17**	**161**

¥ Figures do not cover drones or miscellaneous craft listed in previous tables
† Transferred – hulls may be considerably older
§ Many of these were built over a large number of years and actual dates are not readily available. The majority, however, are in need of replacement.

9.5 South America & Caribbean

The South American and Caribbean region is the one region in the world where the mine threat appears to have had virtually no impact at all. This is surprising in view of the fact that most navies in the region do possess reasonable mine inventories.

There has been no new MCMV construction undertaken for the navies of this region since before 1972. Some navies have acquired second–hand units, particularly Cuba which acquired a number of vessels from the Soviet Union. The latest navy to acquire second–hand units is Uruguay, which acquired four ex–East German Navy *'Kondor'* class vessels.

Of the various navies in the region, the only one likely to have difficulty in sourcing for replacement vessels is Cuba. Of the other Latin American navies all have, in the past, suffered from budgetary restrictions. This situation is now easing, and with the problem of rampant inflation being more easily controlled, there is a possibility that some navies may seek to modernise and strengthen their MCM capability. However, most navies have rather more urgent priorities than upgrading MCM.

Major surface units are desperately in need of modernisation and upgrading, and for the next few years this will top the list of priorities. Possibly not until around the 2000–2005 timescale will the navies of this region seek to address the problem of MCM. When this occurs it is likely that France and Spain will be the favoured potential suppliers, with Germany, Italy and the UK following very close behind.

The region's entire MCM force is desperately in need of replacement, and if these navies are to ensure free movement of shipping in their domain, then there is an urgent need for each to develop a major MCM capability.

TABLE 9.6 MCMV AGE 1972-96[¥] SOUTH AMERICA & CARIBBEAN

Region/Country	1992-96 0-5 years	1987-91 6-10 years	1982-86 11-15 years	1977-81 16-20 years	1972-76 21-25 years	Pre-1972 over 25 years	TOTAL
Argentina						6	6
Brazil					4	2	6
Cuba			7	9			16
Uruguay	4[†]						4
TOTAL	**4**	**0**	**7**	**9**	**4**	**8**	**32**

¥ Figures do not cover drones or miscellaneous craft listed in previous tables
† Transferred – hulls may be considerably older

9.6 North America & Africa

In North America, the United States is nearing the end of a major programme to upgrade its surface MCMV fleet. When this is complete it will possess a modern fleet of 22 vessels which will not require a major upgrading until about the year 2005–2010. At that time any refurbishment or installation of new equipment will most likely be undertaken by American companies.

Across the border to the north, Canada has just begun to commission a new class of MCDVs. These vessels have far ranging capabilities which will meet most of Canada's requirements for the foreseeable future. Again, major mid-life upgrades for these vessels will probably not be required until around 2010–2015.

In Africa only two countries possess a major MCM capability – Nigeria and South Africa. Nigeria does not generally manage its vessels with quite the same degree of care that exists in Western navies. The operational state of the two Italian-built *'Lerici'* class is therefore in some doubt. Nevertheless, these two units will require modernising in about five years time – if they remain in a state worth modernising.

South Africa has maintained its force of MCMVs extremely well in spite of the arms embargo which existed for many years. Although the vessels are well overage, they have been continually refitted and modernised over the years, and are being given a major life extension upgrade to enable them to remain serviceable for up to another 20 years. This has resulted in the vessels being virtually rebuilt. However, at the beginning of the next century South Africa must begin to define a requirement for their replacement and this must begin to take place around 2010.

The only two other countries operating vessels which have had an MCM capability are Angola and Ethiopia. It is highly unlikely that either of these countries will consider replacing current units with new build, and second-hand acquisition will be merely replacement hulls, without necessarily any MCM capability.

TABLE 9.7 MCMV AGE 1972–96[¥]
NORTH AMERICA & AFRICA

Region/Country	**1992–96**	**1987–91**	**1982–86**	**1977–81**	**1972–76**	**Pre–1972**	**TOTAL**
	0–5 years	6–10 years	11–15 years	16–20 years	21–25 years	over 25 years	
Angola		2[†]					**2**
Ethiopia		2[†]					**2**
Canada	4						**4**
Nigeria		2					**2**
USA	14	8					**22**
South Africa				4		4	**8**
TOTAL	**18**	**14**		**4**		**4**	**40**

¥ Figures do not cover drones or miscellaneous craft listed in previous tables

† Transferred – hulls may be considerably older

TABLE 9.8
MCMV CONSTRUCTION STATUS 1996

Country	Class	Name/Number	Builder	Status	Remarks
EUROPE & SCANDINAVIA					
Denmark	SAV	MRD 3/6	Danyard	Building	3/4th of 6
France	Tripartite	Sagittaire (M 650)	Lorient DY	Completed	10th of 10
Germany	Frankenthal	Sulzbach-Rozenburg	Lurssen	Completed	10th of 10
Italy	Gaeta	Chioggia (M 5560)	Intermarine	Completed	7th of 8
		Rimini (M5561)	Intermarine	Completed	8th of 8
Norway	Alta	Otra (M 351)	Kvaerner	Completed	2nd of 5
		Rauma (M 352)	Kvaerner	Completed	3rd of 5
		Orkla (M 353)	Kvaerner	Building	4th of 5
		Glomma (M354)	Kvaerner	Building	5th of 5
Spain	CME	Unnamed (M 51)	Bazan	Building	1st of 4
Sweden	Styrso (YSB)	Styrso (M 11)	KKV	Completed	1st of 4
		Sparo	KKV	Building	2nd of 4
		Skafto	KKV	Building	3rd of 4
		Sturko	KKV	Building	4th of 4
United Kingdom	Sandown	Penzance	Vosper	Building	6th of 12
		Pembroke	Vosper	Building	7th of 12
MIDDLE EAST, GULF, NORTH AFRICA & WESTERN ASIA					
Bangladesh	T 43	M 92	Guangzhou	Building	1st of 3
		M 93	Guangzhou	Building	2nd of 3
		M 94	Guangzhou	Building	3rd of 3
Egypt	Swiftship	CMH 3	Swiftships	Completed	3rd of 3
Pakistan	Tripartite	Muhafiz (167)	Lorient	Completed	2nd of 3
		Mahmood (168)	Karachi	Building	3rd of 3
Saudi Arabia	Sandown	Al Kharj	Vosper	Completed	3rd of 3
ASIA, PACIFIC & AUSTRALASIA					
Australia	Huon	Huon (M82)	ADI	Building	1st of 6
		Hawkesbury	ADI	Building	2nd of 6
Japan	Uwajima	Job No 379 (679)	Hitachi	Completed	8th of 9
		Job No 380 (680)	Nippon	Completed	9th of 9
	MSC 07	MSC 681		Building	1st of 3
		MSC 682		Building	2nd of 3
NORTH AMERICA					
Canada	Kingston	Glace Bay	St John SB	Completed	2nd of 12
		Nanaimo	St John SB	Completed	3rd of 12
		Edmonton	St John SB	Completed	4th of 12
		Shawinigan	St John SB	Building	5th of 12
		Whitehorse	St John SB	Building	6th of 12
		Yellowknife	St John SB	Building	7th of 12
USA	Osprey	Robin (MHC 54)	Avondale	Completed	4th of 12
		Kingfisher (MHC 56)	Avondale	Completed	6th of 12
		Cormorant (MHC 57)	Avondale	Building	7th of 12

TABLE 9.9 PLANNED AND PROJECTED SUBMARINE CONSTRUCTION 1996-2005

Country/Status	Building	On Order	Planned/Projected	Total
EUROPE & SCANDINAVIA				
Belgium		4		**4**
Denmark	4		4	**8***
Germany		2		**2**
Greece			3	**3**
Italy			6	**6**
Norway	2			**2**
Portugal			?	**?**
Poland			?	**?**
Spain	1	3	4	**8**
Sweden	3			**3**
Turkey			6	**6**
UK	2	5		**7**
TOTAL	**12**	**14**	**23+**	**49+**

* slave minesearchers

Country/Status	Building	On Order	Planned/Projected	Total
MIDDLE EAST, GULF, NORTH AFRICA & WESTERN ASIA				
Bangladesh	3			**3**
India			10	**10**
Kuwait			3	**3**
Pakistan	1			**1**
Saudi Arabia			3	**3**
TOTAL	**4**		**16**	**20**
ASIA, PACIFIC & AUSTRALASIA				
Australia	2	4		**6**
Japan	2	1		**3**
Korea (South)			7	**7**
Myanmar (Burma)			2	**2**
Taiwan			12	**12**
Thailand			2	**2**
TOTAL	**4**	**5**	**23**	**32**
NORTH AMERICA & AFRICA				
Canada	3	5		**8**
USA	4			4
TOTAL	**7**	**5**		**12**
GRAND TOTAL	**27**	**24**	**62+**	**113+**

The Italian Gaeta class minehunter/minesweeper Alghero *(H M Steele)*

SECTION 10 - CONCLUSIONS

Although at the present time the market for MCMVs remains fairly static, with a total of 27 hulls under construction and 24 on order, this situation is not likely to continue.

Over the next 5 to 10 years, if all planned programmes and projections are fulfilled, then overall MCMV construction is likely to increase from the present 51 hulls to around 60. Beyond 2005 this figure is likely to increase even more as the number of obsolete hulls requiring replacement grows even greater. At present the number of hulls requiring replacement is in the order of 150. This means that there is a shortfall in replacement of about 100 units. However, it has to be born in mind that of these 100 units, many are operational in navies where there is little likelihood of any new MCMV programmes being formulated in the near future. The number of hulls not likely to be replaced on a one for one basis is in the region of 55. This leaves an overall figure of about 45 hulls requiring replacement and for which, at present, no requirement has been formulated.

In the longer term (up to 2010) the number of hulls requiring replacement will rise rapidly to around 200 units - assuming that the majority of the 45 hulls currently deemed as needing replacement will have been allocated programmes. Of these 200 some 50 are unlikely to be replaced.

Overall the future prospect for MCMV replacement looks good, with a potential for some 200 hulls being required up to about the year 2010. In the folloiwng 10–15 years this figure will increase to around 250 units, of which some 80 are unlikely to be replaced, leaving a total requirement for around 150–170 hulls between 2010 and 2025. Many of these, however, will be for China and Russia, which will leave little opportunity for outside supply due to the indigenous capability of these two countries.

However, these projections all assume that MCMV strength will be retained at its current level. Between now and the year 2010 the political and strategic situation worldwide could change dramatically, with new and unexpected players such as Chile, the Gulf States, Israel and the Pacific Islands and so on, entering the scene.

Governments and navies will have to assess the need for MCMV forces in the light of the existing political situation - worldwide, regional and national - and the real or potential mine threat which they consider they may have to face. On top of this is the ever present question of funding. Already the projected figures outlined above have taken into account those navies where funding will probably dictate against replacement. But other factors will affect funding allocated to MCMV replacement - namely naval priorities appertaining to other vessels in the country's ORBAT and the possible need for their replacement, and a government's perception of a decrease in the strategic threat in a particular region leading to a demand for reduced funding for military projects.

As outlined earlier in this report, the one region apart from Europe which is likely to figure in new MCMV programmes is Asia and the Pacific. Here, the main question revolves around China and its political stance in the coming years. This has far reaching implications in this region. Many countries in the region are currently expanding their naval forces, and MCM is high on the list of priorities, including the acquisition of new MCMVs. The region will also see growth in modernisation and upgrading of MCM assets, and a primary contender for any MCM requirement will be Australia. However, it will face stiff competition from France, Germany, Sweden, and the UK in bidding for any programmes which may be announced.

The next major region with a major potential for an expansion in MCMV forces, is the Gulf. In spite of the reality of the mine threat experienced in this region in recent years, little interest has be shown so far in acquiring MCMVs.

For refurbishment and modernisation India, Pakistan and Saudi Arabia will be the main markets. India, however, is fast developing an effective electronics industry and will make strenuous efforts to become self-sufficient in this field - possibly initially through licence agreements. However, by about 2010 the country should be able to build and outfit an MCMV solely from indigenous design and production. Pakistan and Saudi Arabia will continue to rely on the West for the support of their MCM forces, and here it will be companies in France and the UK who will be prime contenders for any MCM programmes.

In Europe and Scandinavia the main market for the next 5 to 10 years will be in the modernisation and upgrading of MCM assets. There may be an overall reduction in the number of operational units, but towards the end of the first decade of the next century programmes will begin to be initiated for the replacement of obsolete hulls. Any programmes which are initiated will most probably be supplied by domestic companies. Here, however, the situation is clouded by a number of mergers between electronics giants, and probable sources of supply will largely depend on who finally merges with who. The situation in Europe is now reaching a stage where, due to cutbacks in force strength, and the reassessment of national and international needs, overall MCMV strength in the region is starting to diminish. This in turn may lead to further rationalisation in the highly developed yards specialising in MCV construction.

In the North American and African regions there will be little requirement for new construction in the immediate future - although modernisation and upgrading will need to be implemented in a number of units towards the year 2010.

Finally South America and the Caribbean. Here the situation remains decidedly gloomy both for new construction and for upgrading and modernisation. The only country which currently appears to be willing to devote resources towards developing and maintaining an MCM force is Brazil. But all the countries in this region suffer from severe financial restrictions, and MCM does not portray prestige in the same way that major surface warships can. This will be a difficulty market for anyone willing to venture into it, and will probably remain solely in the domain of those European countries who have already supplied military equipment in the past.

From the tables in this report it can be deduced that the trend, therefore, is likely to be of upward growth in new construction towards the time span 2005–2010, with good prospects in the medium term 2000–2005 for mid-life upgrades.

APPENDICES

MANUFACTURERS & CONTRACTORS LIST

A.1 SHIPYARDS

Australia

Australian Defence Industries (ADI) Ltd
PO Box 985
BONDI JUNCTION
New South Wales 2022
Australia
Tel: +61 365 9300
Fax: +61 369 2404

Canada

Saint John Shipbuilding Ltd
PO Box 5111
SAINT JOHN
New Brunswick E2L 4L4
Canada
Tel: +1 506 632 5939
Fax: +1 506 632 5912

Denmark

Danyard A/S
Kragholmen 4
DK 9900 Frederikshavn
Denmark
Tel: +45 9842 2299
Fax: +45 9843 2930

Naval Team Denmark
H.C. Andersens Boulevard 12.3
DK–1553 COPENHAGEN V
Denmark
Tel: +45 33 33 96 94
Fax: +45 33 33 96 54

France

DCN International
19/21 rue de Colonel Pierre–Avia
PO Box 532
F–75725 PARIS CEDEX 15
France
Tel: +33 1 41 08 71 71
Fax: +33 1 41 08 00 27

Germany

Abeking & Rasmussen
PO Box 1160
D–2874 LEMWERDER
Germany
Tel: +49 421 67330

Kroger Werft GmbH & Co KG
PO Box 460
24 754 RENDSBURG
Germany
Tel: +49 4331 951207
Fax: +49 4331 951205

Lurssen Werft
Friedrich–Klippert–Strasse
PO Box 750662
D–28759 BREMEN
Germany
Tel: + 49 421 66040
Fax: +49 421 6604443

Italy

Intermarine
Corso d'Italia No 19
00198 ROME
Italy
Tel: +39 6 841 6113
Fax: +39 6 841 9574

Netherlands

van de Giessen–de Noord
PO Box 13
2950 AA ALBLASSERDAM
The Netherlands
Tel: +31 1859 13844
Fax: +31 1859 12665

Norway

Kvaerner Mandal as
PO Box 283
N 4501 MANDAL
Norway
Tel: + 47 38 27 92 00
Fax: +47 38 26 03 88

Russia

Rosvooruzhenie
18/1 Ovchinnikovskaya Nab
MOSCOW 113324
Russia

Spain

Empresa Nacional Bazan de Construcciones
Navales Militares SA
Paseo de Castellana 55
PO Box 90
E–28046 MADRID
Spain
Tel: +34 1 44 15 100/44 16 100
Fax: +34 1 44 15 50 90

Sweden

Karlskronavarvet
S 371 82 Karlskrona
Sweden
Tel: +46 455 334100
Fax: +46 455 17934

United Kingdom

Vosper Thornycroft (UK) Ltd
Victoria Road
Woolston
SOUTHAMPTON
Hampshire SO19 9RR
UK
Tel: +44 1703 445144

United States of America

Avondale Shipyard
PO Box 50280
NEW ORLEANS
Louisiana 70150–0280
USA
Tel: +1 504 436 2121

Swiftships Inc
PO Box 1908
MORGAN CITY
Louisiana 70380
USA
Tel: +1 504 384 1700
Fax: +1 504 384 0914

A.2 COMMAND AND TRACKING SYSTEMS

Australia

Nautronix Ltd
10 Marine Terrace
FREMANTLE
Western Australia
Australia
Tel: +61 9 430 5900
Fax: +619 430 5901

France

Thomson Marconi Sonar Ltd
525 route des Dolines
BP 157
F-06903 SOPHIA-ANTIPOLIS CEDEX
France
Tel: +33 92 96 45 24
Fax: +33 92 96 40 19

Germany

STN ATLAS Elektronik GmbH
Sebaldsbrücke Heerstrasse 235
PO Box 44 85 45
D-2800 BREMEN 44
Germany
Tel: +49 421 45 70
Fax: +49 421 457 2900

Italy

GF Galileo SMA Srl
Via del Ferrone 5
I-50124 FLORENCE
Italy
Tel: +39 55 27501
Fax: +39 55 714934

Datamat Ingegneria Dei Sistemi SpA
Via Laurentina 760
00143 ROME
Italy
Tel: +39 586 858064
Fax: +39 6 502 0775

Norway

Simrad Subsea A/S
PO Box 111
N-3191 HORTEN
Norway
Tel: +47 33034000

Sweden

CelsiusTech AB
S-175 88 JÄRFÄLLA
Sweden
Tel: +46 758 100 00
Fax: +46 758 322 44

United Kingdom

GEC Marconi S3I
Combat Systems Division
Station Road
ADDLESTONE
Surrey KT15 2PW
UK
Tel: +44 1932 847282
Fax: +44 1932 824723

Kelvin Hughes
New North Road
Hainault
ILFORD
Essex IG6 2UR
UK
Tel: +44 181 500 1020
Fax: +44 181 500 0837

Racal Positioning Systems
Burlington House
118 Burlington Road
NEW MALDEN
Surrey KT3 4NR
UK
Tel: +44 181 942 2464

United States of America

Magnavox Advanced Products & Systems
2829 Maricopa Street
TORRANCE
California 90503
USA
Tel: +1 213 618 1200
Fax: +1 213 618 7001

A.3 SONAR

Canada

C-Tech Ltd
PO Box 1960
525 Boundary Road
CORNWALL
Ontario K6H 6N&
Canada
Tel: +1 613 933 7970
Fax: +1 613 933 7977

Simrad Mesotech Systems Ltd
1598 Kebet Way
PORT COQUITLAM
British Columbia V3C 5W9
Canada
Tel: +1 604 464 8144
Fax: +1 604 941 5423

France

Thomson Marconi Sonar Ltd
525 route des Dolines
BP 157
F-06903 SOPHIA-ANTIPOLIS CEDEX
France
Tel: +33 92 96 45 24
Fax: +33 92 96 40 19

Germany

STN ATLAS Elektronik GmbH
Sebaldsbrücke Heerstrasse 235
PO Box 44 85 45
D-2800 BREMEN 44
Germany
Tel: +49 421 45 70
Fax: +49 421 457 2900

Italy

FIAR
Via Montefeltro 8
I-20156 MILAN
Italy
Tel: +39 2 357901

Norway

Simrad Subsea A/S
PO Box 111
N-3191 HORTEN
Norway
Tel: +47 33034000

United Kingdom

Kelvin Hughes Ltd
New North Road
Hainault
ILFORD
Essex IG6 2UR
UK
Tel: +44 181 500 1020
Fax: +44 181 500 0837

GEC-Marconi Naval Systems Ltd
Sonar Systems Division
Wilkinthroop House
TEMPLECOMBE
Somerset BA8 0DH
UK
Tel: + 44 1963 370551
Fax: +44 1963 442200

Ultra Electronics Ocean Systems
Waverley House
Hampshire Road
Granby Estate
WEYMOUTH
Dorset DT4 9XD
UK
Tel: +44 1305 784738
Fax: +44 1305 777904

United States of America

Klein Associates Inc
11 Klein Drive
SALEM
New Hampshire 03079-1249
USA
Tel: +1 603 893 6131
Fax: +1 603 893 8807

Lockheed Martin
Electronics Park
Building 67
SYRACUSE
New York 13221
USA
Tel: +1 315 456 0123

Raytheon Company
Electronic Systems
Portsmouth Facility
1847 West Main Road
PO Box 360
PORTSMOUTH
Rhode Island 02871–1087
USA
Tel: +1 401 842 3810
Fax: +1 401 842 5214

Simrad Inc
19210 33rd Avenue West
LYNNWOOD
Washington 98036
USA
Tel: +1 206 778 8821
Fax: +1 206 771 7211

Triton Technology Inc
125 Westridge Drive
WATSONVILLE
California 95076
USA
Tel: +1 408 722 7373
Fax: +1 408 722 1405

A.4 UNDERWATER MINE WARFARE VEHICLES

Canada

International Submarine Engineering
Research Ltd
1734 Broadway Street
PORT COQUITLAM
British Columbia V3C 2M8
Canada
Tel: +1 604 942 5223
Fax: +1 604 942 7577

Macdonald Dettwiler
13800 Commerce Parkway
RICHMOND
British Columbia V6V 2J3
Canada
Tel: +1 604 278 3411
Fax: +1 604 278 1285

France

Societe ECA
76 Bd de la Republique
92772 BOULOGNE–BILLANCOURT
Cedex
France
Tel: +33 1 46 10 90 60
Fax: +33 1 46 10 90 70

Thomson Marconi Sonar Ltd
525 route des Dolines
BP 157
F–06903 SOPHIA–ANTIPOLIS CEDEX
France
Tel: +33 92 96 45 24
Fax: +33 92 96 40 19

Germany

STN ATLAS Elektronik GmbH
Sebaldsbrücke Heerstrasse 235
PO Box 44 85 45
D–2800 BREMEN 44
Germany
Tel: +49 421 45 70
Fax: +49 421 457 2900

Italy

Whitehead Alenia
Via di Levante 48
I–57128 LIVORNO
Italy
Tel: +39 586 840 356
Fax: +39 586 854 060

Sweden

Bofors AB
S–691 80 KARLSKOGA
Sweden
Tel: +46 586 81000
Fax: +46 586 85700

United States of America

AlliantTechsystems Inc
6500 Harbour Heights Parkway
MUKILTEO
Washington 98275–4844
USA
Tel: +1 206 356 3000
Fax: +1 206 356 3185

A.5 MINESWEEPING SYSTEMS

Australia

Australian Defence Industries (ADI) Ltd
PO Box 985
BONDI JUNCTION
New South Wales 2022
Australia
Tel: +61 365 9300
Fax: +61 369 2404

Finland

Elesco Oy
Luomannotko 4
PO Box 128
FIN–02201 ESPOO
Finland
Tel: +358 420 8600
Fax: +358 420 8610

Finnyards Ltd
Naulakatu 3
FIN–33100 TAMPERE
Finland
Tel: +358 31 245 0111
Fax: +358 31 213 –188

France

DCN International
19/21 rue de Colonel Pierre–Avia
PO Box 532
F–75725 PARIS CEDEX 15
France
Tel: +33 1 41 08 71 71
Fax: +33 1 41 08 00 27

Germany

Howaldtswerke–Deutsche Werft AG
PO Box 63 09
D–24124 KIEL
Germany
Tel: +49 431 700 3117
Fax: +49 431 700 4211

IBAK Helmut Hunger GmbH
Wehdenweg 122
PO Box 6260
D–24123 KIEL
Germany
Tel: +49 431 727 0223
Fax: +49 431 727 0272

Rheinmetall Industrie GmbH
Pempelfurtstrasse 1
PO Box 1633
D–40836 RATINGEN
Germany
Tel: +49 21 0290 0
Fax: +49 21 0247 3553

Sweden

Bofors Underwater Systems AB
PO Box 627
S–261 02 LANDSKRONA
Sweden
Tel: +46 418 24010
Fax: +46 418 22952

Karlskronavarvet
S 371 82 KARLSKRONA
Sweden
Tel: +46 455 334100
Fax: +46 455 17934

United Kingdom

BAeSEMA
Biwater House
Portsmouth Road
ESHER
Surrey KT10 9SJ
UK
Tel: +44 1372 466660
Fax: +44 1372 466566

United States of America

AlliantTechsystems Inc
6500 Harbour Heights Parkway
MUKILTEO
Washington 98275–4844
USA
Tel: +1 206 356 3000
Fax: +1 206 356 3185

AlliedSignal Inc
Ocean Systems
15825 Roxford Street
SYLMAR
California 91342–3597
USA
Tel: +1 818 367 0111
Fax: +1 818 367 0403

A.6 PROPULSION

France

SEMT Pielstick
2 quai de Seine
F-93202 Saint-Denis
France
Tel: (1) 48 09 76 00
Fax: (1) 42 43 81 02

Germany

MTU Motoren-und Turbinen-Union
Fridrichshafen GmbH
D-88040 FRIEDRICHSHAFEN
Germany
Tel: (7541) 90 0
Fax: (7541) 90 22 47

J.M. Voith GmbH
PO Box 1940
D-7920 HEIDENHEIM
Germany

Schottel Werft
Mainzer-Strasse 99-101
D-540 1 SPAY am RHEIN
Germany
Tel: +49 2628 6137
Fax: +49 2628 61300

Italy

Fincantieri GMT
PO Box 497
34100 TRIESTE
Italy

Sweden

Saab Scania
Scania Division
S-151 87 SODERTALJE
Sweden

United Kingdom

GEC-Alsthom Diesels Ltd
Paxman Works
Hythe Hill
COLCHESTER
Essex CO1 2HW
UK
Tel: (1206) 795151
Fax: (1206) 791238